An introduction to the theory of
Elasticity

R. J. Atkin
N. Fox

**Department of Applied Mathematics
and Computing Science**

University of Sheffield

DOVER PUBLICATIONS, INC
Mineola, New York

Bibliographical Note

This Dover edition, first published in 2005, is an unabridged republication of the edition first published by Longman Group Limited, London, and Longman Inc., New York, 1980.

Library of Congress Cataloging-in-Publication Data

Atkin, R. J. (Raymond John)
  An introduction to the theory of elasticity / R. J. Atkin and N. Fox.
    p. cm.
Originally published: London; New York: Longman, 1980.
Includes bibliographical references and index.
ISBN 0-486-44241-1 (pbk.)
  1. Elasticity. I. Fox, N (Norman), 1936-II. Title.
  QA931.A78 2005
  531'.382—dc22

                                                    2005041433

Manufactured in the United States of America
Dover Publications, Inc., 31 East 2nd Street, Mineola, N.Y. 11501

Contents

5: Anti-plane strain, plane strain, and generalised plane stress

6: Extension, torsion, and bending

7: Elastic waves

Contents

Preface

For many years, continuum mechanics has been recognised as a rich and challenging subject for study, and is firmly established in undergraduate courses in many universities and colleges. Over the last two or three decades, intense research activity in this field has inevitably affected the teaching of the subject. The student's introduction to the once-diverse fields of solid and fluid mechanics has been unified, and as a result, considerable clarity and simplification have been achieved. Our objective here is to place into the hands of second- and third-year undergraduates a text on elasticity, treated from this unified standpoint, which has been prepared while carefully bearing in mind the limited mathematical tools that he is likely to have at his disposal. Much of the material has been given in our own lecture courses over a number of years, and modified in the light of our students' understanding. We have endeavoured to amplify all those sections which have commonly presented difficulties when they have been met with for the first time. Moreover, we have tried to present the material in a manner which will not only be easily assimilated by the student, but will also serve as a foundation for his reading of the many modern advanced texts, both in the finite and infinitesimal theory, which will become necessary should he wish to embark upon postgraduate or research programmes.

In recent years, many lecturers have found that the finite deformation theory of elasticity is a perfectly suitable subject for undergraduate courses, and it is a most desirable background even if the infinitesimal theory is the ultimate goal. Of course, to keep the volume to a reasonable size, we have had to be selective in our choice of subject matter, but we are hopeful that adequate material will be found here that can be presented to students of average ability in an introductory course. If the material is found to be too much for the time available, the syllabus could easily be weighted in favour of finite deformation or the infinitesimal theory according to taste. Chapters 1 to 3 would provide an introduction to finite deformation while some of Chapter 2 and the whole of Chapter 3 could be omitted to provide a course majoring on the infinitesimal theory.

In order to make the text largely self-contained, we have included in Chapter 1 and the early parts of Chapter 2 a discussion of the general principles of continuum mechanics which are necessary for the foundations of elasticity. Where amplification of this material is required, readers are referred to the companion text by Spencer (1980).

A knowledge of vector analysis and calculus, at a level usually reached in introductory university courses on these topics, is assumed. Some familiarity with elementary complex variable theory, Cartesian tensors, and matrix algebra is also required, but sufficient references are given for the student to make good any deficiency he has in these areas.

There has been a trend in recent years towards presenting continuum theories in coordinate-free notation. While this is a simple, elegant tool in the hands of specialists, we have felt that the use of suffix notation in the major part of the text provides the student with an easier transition from the courses in vector analysis that he is likely to have taken. However, direct notation is employed in parts where it provides a clear alternative to suffixes, and where the student is more likely to recognise standard theorems from matrix algebra. The basic theory is derived using Cartesian coordinates, although in applications a limited use of cylindrical and spherical polars is made.

In our presentation of finite deformation elasticity we have, wherever possible, maintained a close link with experiment. We believe that a student's appreciation of, and motivation in, the subject is enhanced if he realizes that the novel effects arising from the theory can not only be observed with relatively simple apparatus, but also used in obtaining quantitative information about the strain-energy function for the material. For this reason we have analysed in detail some non-homogeneous deformations, as well as the simple homogeneous ones, and we have described the associated experiments in which this analysis is used.

When studying the infinitesimal theory of elasticity we have found that there is often confusion in the student's mind concerning the sense in which the deformation is to be regarded as small, and the distinction, if any, which needs to be made between the undeformed and deformed configurations. To clarify these issues we have presented the basic equations in material coordinates so that the displacement and stress are vector and tensor fields defined over the reference configuration. Then, having solved for the displacement field, the deformed configuration can be constructed by superimposing this field on the reference configuration. Some of the worked and unworked examples should clarify potential difficulties in this direction. One price that must be paid in this approach is that the idea of Piola stresses must be introduced. On the whole, we felt that the extra work involved is amply repaid by the gain in understanding of the approximation procedure, and the application of boundary conditions. Moreover, once the equations have been derived, the Piola stresses may be identified with the Cauchy stresses to the degree of approximation used, and so for anyone considering only the applications, a knowledge of the usual Cauchy stress is adequate.

Throughout our presentation we have been greatly influenced by, and have drawn freely from, the articles by Truesdell and Toupin (1960), Truesdell and Noll (1965), and Gurtin (1972). We hope that students will be stimulated to read these major reference books for themselves, as well as the many other volumes listed in our selection of recommendations for further reading (p. 241).

Deformation and stress

In our discussion of the macroscopic behaviour of materials we
disregard their microscopic structure. We think of the material as
being continuously distributed throughout some region of space. At
any instant of time, every point in the region is the location of
what we refer to as a particle of the material. In this chapter we
discuss how the position of each particle may be specified at each
instant, and we introduce certain measures of the change of shape
and size of infinitesimal elements of the material. These measures
are known as strains, and they are used later in the derivation of
the equations of elasticity. We also consider the nature of the
forces acting on arbitrary portions of the body and this leads us
into the concept of stress.

1.1   Motion. Material and spatial coordinates

We wish to discuss the mechanics of bodies composed of various
materials. We idealize the concept of a *body* by supposing that it
is composed of a set of *particles* such that, at each instant of time
t, each particle of the set is assigned to a unique point of a closed
region $\mathscr{C}_t$ of three-dimensional Euclidean space, and that each
point of $\mathscr{C}_t$ is occupied by just one particle. We call $\mathscr{C}_t$ the
configuration of the body at time t.

To describe the *motion* of the body, that is, to specify the
position of each particle at each instant, we require some convenient method of labelling the particles. To do this, we select one
particular configuration $\mathscr{C}$ and call this the reference configuration. The set of coordinates (X_1, X_2, X_3), or position vector $\mathbf{X}$,
referred to fixed Cartesian axes, of a point of $\mathscr{C}$ uniquely determines a particle of the body and may be regarded as a label by
which the particle can be identified for all time. We often refer to
such a particle as the particle $\mathbf{X}$. In choosing $\mathscr{C}$ we are not restricted to those configurations occupied by the body during its

actual motion, although it is often convenient to take $\mathscr{C}$ to be the configuration $\mathscr{C}_0$ occupied by the body at some instant which is taken as the origin of the time scale t.

The motion of the body may now be described by specifying the position $\mathbf{x}$ of the particle $\mathbf{X}$ at time t in the form of an equation

$$\mathbf{x} = \chi(\mathbf{X}, t) \tag{1.1.1}$$

(see Fig. 1.1) or, in component form,

$$x_1 = \chi_1(X_1, X_2, X_3, t), \quad x_2 = \chi_2(X_1, X_2, X_3, t),$$
$$x_3 = \chi_3(X_1, X_2, X_3, t) \tag{1.1.2}$$

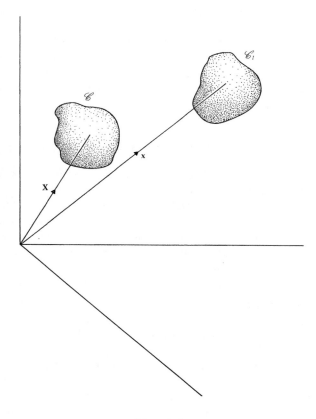

Fig. 1.1

and we assume that the functions χ_1, χ_2, and χ_3 are differentiable with respect to X_1, X_2, X_3, and t as many times as required.

Sometimes we wish to consider only two configurations of the body, an initial configuration and a final configuration. We refer to the mapping from the initial to the final configuration as a *deformation* of the body. The motion of the body may be regarded as a one-parameter sequence of deformations.

We assume that the Jacobian

$$J = \det(\partial\chi_i/\partial X_A), \qquad i, A = 1, 2, 3 \tag{1.1.3}$$

exists at each point of $\mathscr{C}_t$, and that

$$J > 0 \tag{1.1.4}$$

The physical significance of these assumptions is that the material of the body cannot penetrate itself, and that material occupying a finite non-zero volume in $\mathscr{C}$ cannot be compressed to a point or expanded to infinite volume during the motion (see Example 1.3.2).

Mathematically (1.1.4) implies that (1.1.1) has the unique inverse

$$\mathbf{X} = \chi^{-1}(\mathbf{x}, t) \tag{1.1.5}$$

Now at the current time t the position of a typical particle P is given by its Cartesian coordinates (x_1, x_2, x_3), but, as mentioned above, P continues to be identified by the coordinates (X_1, X_2, X_3) which denoted its position in $\mathscr{C}$. The coordinates (X_1, X_2, X_3) are known as *material (or Lagrangian) coordinates* since distinct sets of these coordinates refer to distinct material particles. The coordinates (x_1, x_2, x_3) are known as *spatial (or Eulerian) coordinates* since distinct sets refer to distinct points of space. The values of $\mathbf{x}$ given by equation (1.1.1) for a fixed value of $\mathbf{X}$ are those points of space occupied by the particle $\mathbf{X}$ during the motion. Conversely, the values of $\mathbf{X}$ given by equation (1.1.5) for a fixed value of $\mathbf{x}$ identify the particles $\mathbf{X}$ passing through the point $\mathbf{x}$ during the motion.

From now on, when upper- or lower-case letters are used as suffixes, they are understood generally to range over 1, 2, and 3. Usually upper-case suffixes refer to material coordinates, lower-case to spatial coordinates and repetition of any suffix implies summation over the range. For example, we write x_i for (x_1, x_2, x_3), X_A for (X_1, X_2, X_3) and $x_i x_i$ denotes $x_1^2 + x_2^2 + x_3^2$.

When a quantity is defined at each point of the body at each instant of time, we may express this quantity as a function of X_A and t or of x_i and t. If X_A and t are regarded as the independent variables then the function is said to be a *material description* of the quantity; if x_i and t are used then the corresponding function is said to be a *spatial description*. One description is easily transformed into the other using (1.1.1) or (1.1.5). The material description $\Psi(\mathbf{X}, t)$ has a corresponding spatial description $\psi(\mathbf{x}, t)$ related by

$$\Psi(\boldsymbol{\chi}^{-1}(\mathbf{x}, t), t) = \psi(\mathbf{x}, t) \tag{1.1.6}$$

or

$$\psi(\boldsymbol{\chi}(\mathbf{X}, t), t) = \Psi(\mathbf{X}, t) \tag{1.1.7}$$

To avoid the use of a cumbersome notation and the introduction of a large number of symbols, we usually omit explicit mention of the independent variables, and also use a common symbol for a particular quantity and regard it as denoting sometimes a function of X_A and t, and sometimes the associated function of x_i and t. The following convention for partial differentiation should avoid any confusion.

Let u be the common symbol used to represent a quantity with the material description Ψ and spatial description ψ (these may be scalar-, vector-, or tensor-valued functions) as related by (1.1.6) and (1.1.7). We adopt the following notation for the various partial derivatives:

$$u_{,K} = \frac{\partial \Psi}{\partial X_K}(\mathbf{X}, t), \qquad \frac{\mathrm{D}u}{\mathrm{D}t} = \frac{\partial \Psi}{\partial t}(\mathbf{X}, t) \tag{1.1.8}$$

$$u_{,i} = \frac{\partial \psi}{\partial x_i}(\mathbf{x}, t), \qquad \frac{\partial u}{\partial t} = \frac{\partial \psi}{\partial t}(\mathbf{x}, t) \tag{1.1.9}$$

Example 1.1.1

Write down equations describing the motion of a rigid body moving with constant velocity V in the 1-direction using the configuration $\mathscr{C}_0$ of the body at time $t = 0$ as reference configuration.

If the material description of the temperature u in the body is aX_1, where a is a constant, find the spatial description.

Find also $\mathrm{D}u/\mathrm{D}t$ and $\partial u/\partial t$, and interpret physically.

Using $\mathscr{C}_0$ as reference configuration, the motion is given by

$$x_1 = X_1 + Vt, \quad x_2 = X_2, \quad x_3 = X_3$$

The material and spatial descriptions of the temperature field are

$$u = aX_1 = a(x_1 - Vt)$$

Hence,

$$\frac{Du}{Dt} = 0, \qquad \frac{\partial u}{\partial t} = -aV$$

The coordinates X_A refer to given particles of the body so the temperature $u = aX_1$ remains fixed at each particle. The time derivative Du/Dt measures the time rate of increase of u at fixed X_A, and this is therefore zero. On the other hand, at a fixed point x_i of space the temperature varies as the body passes through and its time rate of increase is $-aV$.

These two time derivatives are of great importance in continuum mechanics and they are discussed in more detail in the next section.

1.2   The material time derivative

Suppose that a certain quantity is defined over the body, and we wish to know its time rate of change as would be recorded at a given particle **X** during the motion. This means that we must calculate the partial derivative, with respect to time, of the material description Ψ of the quantity, keeping **X** fixed. In other words we calculate $\partial\Psi(\mathbf{X}, t)/\partial t$. This quantity is known as a *material time derivative*. We may also calculate the material time derivative from the spatial description ψ. Using the chain rule of partial differentiation, we see from (1.1.7) that

$$\frac{\partial\Psi}{\partial t}(\mathbf{X}, t) = \frac{\partial\psi}{\partial t}(\mathbf{x}, t) + \frac{\partial\chi_i}{\partial t}(\mathbf{X}, t)\frac{\partial\psi}{\partial x_i}(\mathbf{x}, t) \qquad (1.2.1)$$

remembering, of course, that repeated suffixes imply summation over 1, 2, and 3.

Consider now a given particle $\mathbf{X}_0$. Its position in space at time t is

$$\mathbf{x} = \boldsymbol{\chi}(\mathbf{X}_0, t)$$

and so its velocity and acceleration are

$$\frac{d\boldsymbol{\chi}}{dt}(\mathbf{X}_0, t) \quad \text{and} \quad \frac{d^2\boldsymbol{\chi}}{dt^2}(\mathbf{X}_0, t)$$

respectively. We therefore define the *velocity field* for the particles of the body to be the material time derivative $\partial\boldsymbol{\chi}(\mathbf{X}, t)/\partial t$, and use the common symbol $\mathbf{v}$ to denote its material or spatial description:

$$\mathbf{v} = \frac{\partial\boldsymbol{\chi}}{\partial t}(\mathbf{X}, t) = \frac{D\mathbf{x}}{Dt} \tag{1.2.2}$$

Likewise we define the *acceleration field* f to be the material time derivative of $\mathbf{v}$:

$$\mathbf{f} = \frac{D\mathbf{v}}{Dt} \tag{1.2.3}$$

Moreover, in view of (1.2.1) the material time derivative of u has the equivalent forms

$$\frac{Du}{Dt} = \frac{\partial u}{\partial t} + v_i u_{,i} \tag{1.2.4}$$

In particular, the acceleration (1.2.3) may be written as

$$\mathbf{f} = \frac{D\mathbf{v}}{Dt} = \frac{\partial\mathbf{v}}{\partial t} + (\mathbf{v} \cdot \boldsymbol{\nabla})\mathbf{v} \tag{1.2.5}$$

where the operator $\boldsymbol{\nabla}$ is defined relative to the coordinates x_i, that is,

$$\boldsymbol{\nabla} = \left(\frac{\partial}{\partial x_1}, \frac{\partial}{\partial x_2}, \frac{\partial}{\partial x_3}\right) \tag{1.2.6}$$

In suffix notation, (1.2.5) becomes

$$f_i = \frac{Dv_i}{Dt} = \frac{\partial v_i}{\partial t} + v_j v_{i,j} \tag{1.2.7}$$

Example 1.2.1

Find the value of J and the material and spatial descriptions of the velocity field for the motion given by

$$x_1 = X_1 + \alpha t X_2^2, \quad x_2 = (1 + \beta t)X_2, \quad x_3 = X_3, \quad t \geq 0 \tag{1.2.8}$$

where α and β ($\geqslant 0$) are constants. Calculate $\partial v/\partial t$, $(\mathbf{v}\cdot\boldsymbol{\nabla})\mathbf{v}$, and $D\mathbf{v}/Dt$. Verify the relation (1.2.4).

For this motion,

$$J = \det\left(\frac{\partial x_i}{\partial X_K}\right) = \begin{vmatrix} 1 & 2\alpha t X_2 & 0 \\ 0 & 1+\beta t & 0 \\ 0 & 0 & 1 \end{vmatrix} = 1+\beta t > 0$$

Using (1.2.2),

$$\mathbf{v} = (\alpha X_2^2, \beta X_2, 0) \qquad (1.2.9)$$

which is the material description of the velocity field. Equations (1.2.8) may be inverted to give

$$X_1 = x_1 - \frac{\alpha t x_2^2}{(1+\beta t)^2}, \quad X_2 = \frac{x_2}{1+\beta t}, \quad X_3 = x_3$$

and so the spatial description of the velocity field may be written

$$\mathbf{v} = \left(\frac{\alpha x_2^2}{(1+\beta t)^2}, \frac{\beta x_2}{1+\beta t}, 0\right)$$

Hence,

$$\frac{\partial \mathbf{v}}{\partial t} = \left(-\frac{2\alpha\beta x_2^2}{(1+\beta t)^3}, -\frac{\beta^2 x_2}{(1+\beta t)^2}, 0\right),$$

and

$$(\mathbf{v}\cdot\boldsymbol{\nabla})\mathbf{v} = \left(\frac{\alpha x_2^2}{(1+\beta t)^2}\frac{\partial}{\partial x_1} + \frac{\beta x_2}{1+\beta t}\frac{\partial}{\partial x_2}\right)\left(\frac{\alpha x_2^2}{(1+\beta t)^2}, \frac{\beta x_2}{1+\beta t}, 0\right)$$

$$= \left(\frac{2\alpha\beta x_2^2}{(1+\beta t)^3}, \frac{\beta^2 x_2}{(1+\beta t)^2}, 0\right)$$

Also, from (1.2.9),

$$\frac{D\mathbf{v}}{Dt} = (0, 0, 0)$$

and so (1.2.4) is verified.

1.3   The deformation-gradient tensor

We have discussed how the motion of a body may be described. In this section we analyse the deformation of infinitesimal elements of the body which results from this motion.

Suppose that $\mathscr{C}$ coincides with the initial configuration $\mathscr{C}_0$, and that two neighbouring particles P and Q have positions $\mathbf{X}$ and $\mathbf{X} + d\mathbf{X}$ in $\mathscr{C}$. Then at time t their positions in $\mathscr{C}_t$ are $\mathbf{x}$ and $\mathbf{x} + d\mathbf{x}$, where

$$\mathbf{x} = \chi(\mathbf{X}, t), \quad \mathbf{x} + d\mathbf{x} = \chi(\mathbf{X} + d\mathbf{X}, t) \tag{1.3.1}$$

and the components of the total differential $d\mathbf{x}$ are given in terms of the components of $d\mathbf{X}$ and the partial derivatives of χ by

$$dx_i = \frac{\partial \chi_i}{\partial X_A}(\mathbf{X}, t)\, dX_A = x_{i,A}\, dX_A \tag{1.3.2}$$

The quantities $x_{i,A}$ are known as the *deformation gradients*. They are the components of a second-order tensor known as the *deformation-gradient tensor* which we denote by $\mathbf{F}$. (Readers unfamiliar with Cartesian tensors should consult, for example, Spencer (1980) Ch. 3.)

Example 1.3.1

Show that when the coordinate axes are rotated about the origin, and the coordinates x_i and X_A are transformed into

$$x_i' = l_{ij}x_j, \qquad X_A' = l_{AB}X_B$$

where l_{ij} are the direction cosines of the coordinate transformation, the components F_{iA} of $\mathbf{F}$ are transformed into

$$F_{iA}' = \frac{\partial x_i'}{\partial X_A'} = l_{ij}l_{AB}F_{jB}$$

(This proves that the deformation gradients are the components of a second-order tensor.)

Example 1.3.2

Show that the assumption (1.1.4), $J > 0$, implies that the material of the body cannot penetrate itself, and that material occupying a finite non-zero volume in $\mathscr{C}$ cannot be compressed to a point or expanded to infinite volume during the motion. Deduce also that a volume element dV_0 in $\mathscr{C}$ deforms into a volume element dV in $\mathscr{C}_t$ where $dV = J\, dV_0$.

Consider three infinitesimal non-coplanar line elements $d\mathbf{X}^{(1)}$, $d\mathbf{X}^{(2)}$, $d\mathbf{X}^{(3)}$ at a point P in $\mathscr{C}$, and suppose that they correspond to line elements $d\mathbf{x}^{(1)}$, $d\mathbf{x}^{(2)}$, $d\mathbf{x}^{(3)}$ in $\mathscr{C}_t$. Then

$$dx_i^{(\alpha)} = x_{i,A}\, dX_A^{(\alpha)}, \qquad \alpha = 1, 2, 3$$

Hence

$$\begin{aligned} \det{(dx_i^{(\alpha)})} &= \det{(x_{i,A}\, dX_A^{(\alpha)})} \\ &= J \det{(dX_A^{(\alpha)})} \end{aligned} \tag{1.3.3}$$

since the determinant of the product of two matrices is equal to the product of their determinants. But

$$\begin{aligned} \det{(dx_i^{(\alpha)})} &= d\mathbf{x}^{(1)} \cdot d\mathbf{x}^{(2)} \times d\mathbf{x}^{(3)} \\ \det{(dX_A^{(\alpha)})} &= d\mathbf{X}^{(1)} \cdot d\mathbf{X}^{(2)} \times d\mathbf{X}^{(3)} \end{aligned} \tag{1.3.4}$$

Now $d\mathbf{X}^{(1)} \cdot d\mathbf{X}^{(2)} \times d\mathbf{X}^{(3)}$ is positive or negative according as to whether $d\mathbf{X}^{(1)}$, $d\mathbf{X}^{(2)}$, $d\mathbf{X}^{(3)}$ are ordered in a right-handed or left-handed sense, and this triple scalar product has magnitude equal to the volume of a parallelepiped, three of whose edges are taken to be $d\mathbf{X}^{(1)}$, $d\mathbf{X}^{(2)}$, $d\mathbf{X}^{(3)}$. A similar result holds for $d\mathbf{x}^{(1)}$, $d\mathbf{x}^{(2)}$, $d\mathbf{x}^{(3)}$. From (1.3.3) and (1.3.4) we see therefore that, if $J > 0$, a right-handed triad cannot deform into a left-handed triad, and vice versa. In other words, one line element cannot penetrate the plane of the other two. Moreover, the volume dV of the parallelepiped determined by $d\mathbf{x}^{(\alpha)}$ cannot become zero or infinite: it is related to the volume dV_0 of the parallelepiped determined by $d\mathbf{X}^{(\alpha)}$ through the relation

$$dV = J\, dV_0 \tag{1.3.5}$$

Example 1.3.3

Show that

$$\frac{DJ}{Dt} = J \operatorname{div} \mathbf{v} \tag{1.3.6}$$

The ε-symbol is defined by

$$\varepsilon_{ABC} = \begin{cases} +1 & \text{if A, B, C is any cyclic permutation of 1, 2, 3} \\ & \quad \text{(i.e. 1, 2, 3; 2, 3, 1; or 3, 1, 2)} \\ -1 & \text{if A, B, C is any anti-cyclic permutation of 1, 2, 3} \\ & \quad \text{(i.e. 2, 1, 3; 1, 3, 2; or 3, 2, 1)} \\ 0 & \text{if any two of A, B, C are equal} \end{cases}$$

Using this symbol, the expansion

$$J = \varepsilon_{ABC} x_{1,A} x_{2,B} x_{3,C} \tag{1.3.7}$$

is easily verified. Hence

$$\frac{DJ}{Dt} = \varepsilon_{ABC} \left\{ \frac{D}{Dt}(x_{1,A}) x_{2,B} x_{3,C} + x_{1,A} \frac{D}{Dt}(x_{2,B}) x_{3,C} \right.$$
$$\left. + x_{1,A} x_{2,B} \frac{D}{Dt}(x_{3,C}) \right\} \tag{1.3.8}$$

Now

$$\frac{D}{Dt}(x_{1,A}) = \frac{\partial^2}{\partial t \, \partial X_A} x_1(\mathbf{X}, t) = \frac{\partial v_1}{\partial X_A}$$
$$= \frac{\partial v_1}{\partial x_j} \frac{\partial x_j}{\partial X_A}$$

using the chain rule of partial differentiation. Since the value of a determinant with two identical rows is zero, and using (1.3.7), we see that

$$\varepsilon_{ABC} \frac{D}{Dt}(x_{1,A}) x_{2,B} x_{3,C} = \frac{\partial v_1}{\partial x_j} \varepsilon_{ABC} x_{j,A} x_{2,B} x_{3,C}$$
$$= \frac{\partial v_1}{\partial x_1} J$$

Similar expressions may be derived for the remaining terms of (1.3.8) and the result (1.3.6) follows.

Example 1.3.4

Show that

$$J \varepsilon_{KLM} = \varepsilon_{ijk} x_{i,K} x_{j,L} x_{k,M} \tag{1.3.9}$$

[*Hint:* From the theory of determinants we know that: (i) any cyclic interchange of the columns of a determinant leaves its value unchanged; (ii) the interchange of any two columns multiplies the value by -1; and (iii) if any two columns are identical the value of the determinant is zero. Now use (1.3.7) and consider separately the three possibilities: K, L, M a cyclic permutation of 1, 2, 3; an anti-cyclic permutation; and at least two of K, L, M equal.]

1.4 Strain tensors

Denoting the deformation gradients $x_{i,A}$ by F_{iA}, equation (1.3.2) may be written

$$dx_i = F_{iA} \, dX_A \qquad (1.4.1)$$

In view of our assumption (1.1.4), the tensor **F** is non-singular, and so permits the unique decompositions

$$\mathbf{F} = \mathbf{RU}, \qquad \mathbf{F} = \mathbf{VR} \qquad (1.4.2)$$

where **U** and **V** are positive-definite symmetric tensors and **R** is proper orthogonal (see the polar decomposition theorem (Spencer (1980) Section 2.5). We remind the reader that a proper orthogonal tensor **R** has the properties

$$\mathbf{R}^{\mathrm{T}}\mathbf{R} = \mathbf{RR}^{\mathrm{T}} = \mathbf{I}, \qquad \det \mathbf{R} = 1 \qquad (1.4.3)$$

where $\mathbf{R}^{\mathrm{T}}$ denotes the transpose of **R**, and **I** denotes the unit tensor. A positive-definite tensor **U** has the property

$$x_i U_{ij} x_j > 0 \qquad (1.4.4)$$

for all non-null vectors **x**.

To see the physical significance of the decomposition (1.4.2), we first write (1.4.1) in the form

$$dx_i = R_{iK} U_{KL} \, dX_L \qquad (1.4.5)$$

or, equivalently,

$$dx_i = R_{iK} \, dy_K, \qquad dy_K = U_{KL} \, dX_L \qquad (1.4.6)$$

In other words, the deformation of line elements $d\mathbf{X}$ into $d\mathbf{x}$, caused by the motion, may be split into two parts. Since **U** is a positive-definite symmetric tensor, there exists a set of axes, known as *principal axes*, referred to which **U** is diagonal; and the diagonal components are the positive principal values U_1, U_2, U_3 of **U**. Equation (1.4.6)$_2$,* referred to these axes, becomes

$$dy_1 = U_1 \, dX_1, \quad dy_2 = U_2 \, dX_2, \quad dy_3 = U_3 \, dX_3 \qquad (1.4.7)$$

In the deformation represented by equations (1.4.7), the ith component of each line element is increased or diminished in magnitude according as $U_i > 1$ or $U_i < 1$. This part of the deformation

---

* The notation (1.4.6)$_2$ refers to the second equation of the set labelled (1.4.6). This convention is adopted throughout the text.

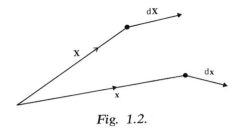

Fig. 1.2.

therefore amounts to a simple stretching or compression in three mutually perpendicular directions. (Of course if $U_i = 1$ the corresponding component of the line element is unchanged.) The values of U_i are known as the *principal stretches*.

Equation $(1.4.6)_1$ describes a rigid-body rotation of the line elements d**y** to d**x**. Hence, the line elements d**X** may be thought of as being first translated from **X** to **x**, then stretched by the tensor **U** as described above, and finally rotated as a rigid body in a manner determined by **R** (see Fig. 1.2). The decomposition $(1.4.2)_2$ may be interpreted in a similar way, although it should be noted that in this case the rotation comes before the stretching. The tensors **U** and **V** are known as the *right* and *left stretching tensors* respectively.

Although the decompositions (1.4.2) provide useful measures of the local stretching of an element of the body as distinct from its rigid-body rotation, the calculation of the tensors **U** and **V** for any but the simplest deformations can be tedious. For this reason we define two more convenient measures of the stretching part of the deformation. We define the *right* and *left Cauchy–Green strain tensors*

$$\mathbf{C} = \mathbf{F}^T \mathbf{F}, \qquad \mathbf{B} = \mathbf{F}\mathbf{F}^T \tag{1.4.8}$$

respectively. Clearly **C** and **B** are symmetric second-order tensors. The tensor **C** is easily related to **U** since, using $(1.4.2)_1$ and $(1.4.3)_1$,

$$\mathbf{C} = \mathbf{U}^T \mathbf{R}^T \mathbf{R} \mathbf{U} = \mathbf{U}^T \mathbf{U} = \mathbf{U}^2 \tag{1.4.9}$$

Similarly, we can show that

$$\mathbf{B} = \mathbf{V}^2 \tag{1.4.10}$$

As can be seen from the definitions (1.4.8), when **F** has been found, the tensors **B** and **C** are easily calculated by matrix multiplication; and, in principle, **U** and **V** can be determined as the unique positive-definite square roots of $\mathbf{C}$ and $\mathbf{B}$. The following example has been constructed so that $\mathbf{C}$ and $\mathbf{B}$ are diagonal. In such cases $\mathbf{U}$ and $\mathbf{V}$ can be found easily.

Example 1.4.1

Find the tensors $\mathbf{F}$, $\mathbf{C}$, $\mathbf{B}$, $\mathbf{U}$, $\mathbf{V}$, and $\mathbf{R}$ for the deformation

$$x_1 = X_1, \quad x_2 = X_2 - \alpha X_3, \quad x_3 = X_3 + \alpha X_2 \tag{1.4.11}$$

where α (>0) is a constant, and interpret the deformation as a sequence of stretches and a rotation.

For this deformation

$$\mathbf{F} = \begin{pmatrix} 1 & 0 & 0 \\ 0 & 1 & -\alpha \\ 0 & \alpha & 1 \end{pmatrix}, \qquad J = 1 + \alpha^2 > 0 \tag{1.4.12}$$

Hence,

$$\mathbf{C} = \mathbf{F}^T \mathbf{F} = \begin{pmatrix} 1 & 0 & 0 \\ 0 & 1 & \alpha \\ 0 & -\alpha & 1 \end{pmatrix} \begin{pmatrix} 1 & 0 & 0 \\ 0 & 1 & -\alpha \\ 0 & \alpha & 1 \end{pmatrix}$$

$$= \begin{pmatrix} 1 & 0 & 0 \\ 0 & 1+\alpha^2 & 0 \\ 0 & 0 & 1+\alpha^2 \end{pmatrix} \tag{1.4.13}$$

and therefore

$$\mathbf{U} = \begin{pmatrix} 1 & 0 & 0 \\ 0 & (1+\alpha^2)^{\frac{1}{2}} & 0 \\ 0 & 0 & (1+\alpha^2)^{\frac{1}{2}} \end{pmatrix} \tag{1.4.14}$$

It can easily be shown likewise that $\mathbf{B} = \mathbf{C}$ and $\mathbf{V} = \mathbf{U}$. We may calculate $\mathbf{R}$ from the relation $\mathbf{R} = \mathbf{F}\mathbf{U}^{-1}$. Thus

$$\mathbf{R} = \begin{pmatrix} 1 & 0 & 0 \\ 0 & 1 & -\alpha \\ 0 & \alpha & 1 \end{pmatrix} \begin{pmatrix} 1 & 0 & 0 \\ 0 & (1+\alpha^2)^{-\frac{1}{2}} & 0 \\ 0 & 0 & (1+\alpha^2)^{-\frac{1}{2}} \end{pmatrix}$$

$$= \begin{pmatrix} 1 & 0 & 0 \\ 0 & (1+\alpha^2)^{-\frac{1}{2}} & -\alpha(1+\alpha^2)^{-\frac{1}{2}} \\ 0 & \alpha(1+\alpha^2)^{-\frac{1}{2}} & (1+\alpha^2)^{-\frac{1}{2}} \end{pmatrix} \tag{1.4.15}$$

Now let $\alpha = \tan\theta$ $(0 < \theta < \tfrac{1}{2}\pi)$, then

$$\mathbf{R} = \begin{pmatrix} 1 & 0 & 0 \\ 0 & \cos\theta & -\sin\theta \\ 0 & \sin\theta & \cos\theta \end{pmatrix}$$

which represents a rotation through an angle $-\theta$ about the 1-axis, using the usual corkscrew convention for the sign of the angle. Thus the deformation may be accomplished by first performing stretches of magnitudes $(1+\alpha^2)^{\frac{1}{2}}$ in the 2- and 3-directions and then a rotation about the 1-axis. Since in this example $\mathbf{B} = \mathbf{C}$ and $\mathbf{V} = \mathbf{U}$, these operations may be reversed in order.

If a portion of the body moves in such a manner that the distances between every pair of particles remain constant, that portion is said to move as a rigid body. For such a motion no stretching of line elements occurs, and so, at each particle of the given portion,

$$\mathbf{B} = \mathbf{C} = \mathbf{U} = \mathbf{V} = \mathbf{I}, \qquad \mathbf{F} = \mathbf{R} \tag{1.4.16}$$

In general, of course, the motion of the body does produce changes in the lengths of line elements, and an analysis of these length changes leads us to an alternative interpretation of $\mathbf{C}$ and $\mathbf{B}$. Suppose that dL and dl denote the lengths of the vector line elements $d\mathbf{X}$ and $d\mathbf{x}$, respectively. Then, using (1.3.2), (1.4.8)$_1$, and the Kronecker delta (defined by $\delta_{KL} = 0$, $K \neq L$; $\delta_{KL} = 1$, $K = L$),

$$\begin{aligned} (dl)^2 - (dL)^2 &= dx_i\, dx_i - dX_K\, dX_K \\ &= x_{i,K}\, x_{i,L}\, dX_K\, dX_L - dX_K\, dX_K \\ &= (C_{KL} - \delta_{KL})\, dX_K\, dX_L \end{aligned} \tag{1.4.17}$$

and so the tensor $\mathbf{C}$ enables us to calculate the difference between the squared elements of length in the reference and current configurations. Alternatively, if we define the inverse deformation gradients, using (1.1.5), as

$$X_{K,i} = \frac{\partial}{\partial x_i}\, \chi_K^{-1}(\mathbf{x}, t) \tag{1.4.18}$$

then, since $X_{K,i} x_{i,A} = \delta_{KA}$ by the chain rule of partial differentiation, it follows from (1.3.2) that

$$dX_K = X_{K,i}\, dx_i \tag{1.4.19}$$

Hence, we may write

$$(dl)^2 - (dL)^2 = dx_i\, dx_i - X_{K,i}X_{K,j}\, dx_i\, dx_j$$

It can easily be verified, using (1.4.8) and the result $(\mathbf{F}^T)^{-1} = (\mathbf{F}^{-1})^T$, that

$$(\mathbf{F}^{-1})^T\mathbf{F}^{-1} = \mathbf{B}^{-1}, \qquad X_{K,i}X_{K,j} = B_{ij}^{-1} \qquad (1.4.20)$$

and therefore

$$(dl)^2 - (dL)^2 = (\delta_{ij} - B_{ij}^{-1})\, dx_i\, dx_j \qquad (1.4.21)$$

The tensor **B** also provides us with a means of calculating the same difference of squared elements of length.

As we have already noted, **B** and **C** are second-order symmetric tensors. Their principal axes and principal values are real, and may be found in the usual manner (Spencer (1980) Sections 2.3 and 9.3). The characteristic equation for **C** is

$$\det(C_{KL} - \lambda\delta_{KL}) = 0$$

that is,

$$\lambda^3 - I_1\lambda^2 + I_2\lambda - I_3 = 0 \qquad (1.4.22)$$

where

$$
\begin{aligned}
I_1 &= C_{KK} = \operatorname{tr}\mathbf{C} \\
I_2 &= \tfrac{1}{2}(C_{KK}C_{LL} - C_{KL}\,C_{KL}) = \tfrac{1}{2}(\operatorname{tr}\mathbf{C})^2 - \tfrac{1}{2}\operatorname{tr}\mathbf{C}^2 \qquad (1.4.23)\\
I_3 &= \det\mathbf{C}
\end{aligned}
$$

and tr denotes the trace. The quantities I_1, I_2, and I_3 are known as the *principal invariants* of **C**.

Example 1.4.2

On rotation of axes, the components of **C** are transformed by the usual tensor transformation law. Show that the values of I_1, I_2, and I_3 are unchanged.

Example 1.4.3

Show that the principal invariants of **B** are the same as those of **C**, but that the principal axes of **B** and **C** do not necessarily coincide.

[*Hint:* Having shown that the invariants are identical, all that is necessary to prove that the principal axes do not always coincide is to find a deformation for which $\mathbf{B} \neq \mathbf{C}$.]

We also note here a useful physical interpretation of I_3. In view of the definitions (1.1.3) and (1.4.8)$_1$,

$$I_3 = \det \mathbf{C} = (\det \mathbf{F})^2 = J^2 \tag{1.4.24}$$

and, if a given set of particles occupies an element of volume dV_0 in $\mathscr{C}$ and dV in $\mathscr{C}_t$, then using (1.3.5),

$$J = dV/dV_0 \tag{1.4.25}$$

Thus, recalling (1.1.4),

$$dV/dV_0 = \sqrt{I_3} \tag{1.4.26}$$

If no volume change occurs during the deformation, the deformation is said to be *isochoric*, and

$$J = 1, \qquad I_3 = 1 \tag{1.4.27}$$

The strain invariants are also of fundamental importance in the constitutive theory of elasticity discussed in Chapter 2. In that chapter we also find the following relation useful:

$$I_2 = I_3 \operatorname{tr} (\mathbf{B}^{-1}) \tag{1.4.28}$$

To prove this, we first note that, from the Cayley–Hamilton theorem (Spencer (1980) Section 2.4) a matrix satisfies its own characteristic equation. Since the principal invariants of $\mathbf{B}$ are identical to those of $\mathbf{C}$, $\mathbf{B}$ must satisfy the equation

$$\mathbf{B}^3 - I_1\mathbf{B}^2 + I_2\mathbf{B} - I_3\mathbf{I} = 0 \tag{1.4.29}$$

Now $\mathbf{B}$ is non-singular, so multiplying (1.4.29) by $\mathbf{B}^{-1}$, we find that

$$\mathbf{B}^2 = I_1\mathbf{B} - I_2\mathbf{I} + I_3\mathbf{B}^{-1}$$

Taking the trace of this equation, we have

$$\operatorname{tr} (\mathbf{B}^2) = I_1^2 - 3I_2 + I_3 \operatorname{tr} (\mathbf{B}^{-1}) \tag{1.4.30}$$

and using (1.4.23)$_2$, (1.4.30) reduces to (1.4.28).

1.5   Homogeneous deformation

A deformation of the form

$$x_i = A_{iK}X_K + a_i \tag{1.5.1}$$

in which **A** and **a** are constants, is known as a *homogeneous deformation*. Clearly **F** = **A** and $J = \det \mathbf{A}$. Particularly simple examples of such deformations are given below.

(i) Dilatation

Consider the deformation

$$x_1 = \alpha X_1, \quad x_2 = \alpha X_2, \quad x_3 = \alpha X_3 \tag{1.5.2}$$

where α is a constant, then

$$\mathbf{F} = \alpha \mathbf{I}, \quad \mathbf{B} = \mathbf{C} = \alpha^2 \mathbf{I}, \quad J = \alpha^3 \tag{1.5.3}$$

and so, to satisfy (1.1.4) we must have $\alpha > 0$. The strain invariants (1.4.23) are easily seen to be

$$I_1 = 3\alpha^2, \quad I_2 = 3\alpha^4, \quad I_3 = \alpha^6 \tag{1.5.4}$$

In view of (1.4.25) we see that if $\alpha > 1$ the deformation represents an expansion; if $\alpha < 1$ the deformation becomes a contraction.

(ii) Simple extension with lateral extension or contraction

Suppose that

$$x_1 = \alpha X_1, \quad x_2 = \beta X_2, \quad x_3 = \beta X_3 \tag{1.5.5}$$

Then

$$\mathbf{F} = \begin{pmatrix} \alpha & 0 & 0 \\ 0 & \beta & 0 \\ 0 & 0 & \beta \end{pmatrix}, \quad J = \alpha\beta^2 \tag{1.5.6}$$

and so $\alpha > 0$. If $\alpha > 1$, the deformation is a uniform extension in the 1-direction; if $\alpha < 1$, the deformation is a uniform contraction in the 1-direction (see Example 1.5.2). If $\beta > 0$ the diagonal terms of **F** are all positive so that **U** = **F** and **R** = **I**; β measures the lateral extension ($\beta > 1$), or contraction ($\beta < 1$), in the 2,3-plane.

If $\beta < 0$ then

$$\mathbf{U} = \begin{pmatrix} \alpha & 0 & 0 \\ 0 & -\beta & 0 \\ 0 & 0 & -\beta \end{pmatrix}, \qquad \mathbf{R} = \begin{pmatrix} 1 & 0 & 0 \\ 0 & -1 & 0 \\ 0 & 0 & -1 \end{pmatrix}$$

In this case $-\beta$ measures the associated lateral extension or contraction, and a rotation through an angle π about the 1-axis is included in the deformation.

If the material is incompressible (see Section 2.1), only isochoric deformations are possible, in which case

$$J = \alpha\beta^2 = 1 \tag{1.5.7}$$

This means that $|\beta|$ is less than, or greater than, unity according as to whether α is greater than, or less than, unity. In other words, an extension in the 1-direction produces a contraction in the lateral directions and vice versa. The strain tensors are found to be

$$\mathbf{B} = \mathbf{C} = \begin{pmatrix} \alpha^2 & 0 & 0 \\ 0 & \beta^2 & 0 \\ 0 & 0 & \beta^2 \end{pmatrix} \tag{1.5.8}$$

and the invariants are

$$I_1 = \alpha^2 + 2\beta^2, \quad I_2 = 2\alpha^2\beta^2 + \beta^4, \quad I_3 = \alpha^2\beta^4 \tag{1.5.9}$$

Example 1.5.1

The previous two deformations are special cases of

$$x_1 = \lambda_1 X_1, \quad x_2 = \lambda_2 X_2, \quad x_3 = \lambda_3 X_3 \tag{1.5.10}$$

where λ_i $(i = 1, 2, 3)$ are constants. Show that, for the deformation (1.5.10) to satisfy $J > 0$, at least one of the λ_i has to be positive. Interpret the deformation geometrically in the case when all the λ_i are positive, and show that, in all cases, the principal invariants are

$$I_1 = \lambda_1^2 + \lambda_2^2 + \lambda_3^2, \quad I_2 = \lambda_1^2\lambda_2^2 + \lambda_2^2\lambda_3^2 + \lambda_3^2\lambda_1^2, \quad I_3 = \lambda_1^2\lambda_2^2\lambda_3^2 \tag{1.5.11}$$

Example 1.5.2

The particles which are at $(L, 0, 0)$ and $(L + D, 0, 0)$ in $\mathscr{C}$ are a distance D apart. Find their distance apart in the deformed configuration when the deformation is given by (1.5.5).

(iii) Simple shear

Consider the deformation

$$x_1 = X_1 + \kappa X_2, \quad x_2 = X_2, \quad x_3 = X_3 \qquad (1.5.12)$$

where κ is a constant. The particles move only in the 1-direction, and their displacement is proportional to their 2-coordinate. This deformation is known as a *simple shear*. Planes parallel to $X_1 = 0$ are rotated about an axis parallel to the 3-axis through an angle $\theta = \tan^{-1} \kappa$, known as the *angle of shear*. The sense of the rotation is indicated in Fig. 1.3. The planes $X_3 = \text{constant}$ are known as the *planes of shear;* the planes $X_2 = \text{constant}$ are called *shearing planes;* and lines parallel to the X_3-axis are known as *axes of shear*.

The deformation-gradient tensor is

$$\mathbf{F} = \begin{pmatrix} 1 & \kappa & 0 \\ 0 & 1 & 0 \\ 0 & 0 & 1 \end{pmatrix} \qquad (1.5.13)$$

and the Cauchy–Green strain tensors are

$$\mathbf{C} = \mathbf{F}^T \mathbf{F} = \begin{pmatrix} 1 & \kappa & 0 \\ \kappa & 1+\kappa^2 & 0 \\ 0 & 0 & 1 \end{pmatrix}, \quad \mathbf{B} = \mathbf{F}\mathbf{F}^T = \begin{pmatrix} 1+\kappa^2 & \kappa & 0 \\ \kappa & 1 & 0 \\ 0 & 0 & 1 \end{pmatrix} \quad (1.5.14)$$

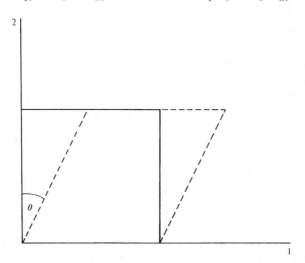

Fig. 1.3

The tensor $\mathbf{B}^{-1}$ may be found using the formula

$$\mathbf{B}^{-1} = \text{adj } \mathbf{B}/\det \mathbf{B} \qquad (1.5.15)$$

where adj $\mathbf{B}$ denotes the adjoint matrix of $\mathbf{B}$. Thus

$$\mathbf{B}^{-1} = \begin{pmatrix} 1 & -\kappa & 0 \\ -\kappa & 1+\kappa^2 & 0 \\ 0 & 0 & 1 \end{pmatrix} \qquad (1.5.16)$$

The strain invariants are

$$I_1 = 3+\kappa^2, \quad I_2 = 3+\kappa^2, \quad I_3 = 1 \qquad (1.5.17)$$

Example 1.5.3

Verify the relation (1.4.28) for this deformation.

Example 1.5.4

Interpret the following deformations geometrically:

(a) $x_1 = \alpha X_1 + \kappa X_2$, $x_2 = X_2$, $x_3 = X_3$
(b) $x_1 = X_1 + \kappa X_2$, $x_2 = X_2 + \kappa X_3$, $x_3 = X_3$

where α and κ are positive constants.

Example 1.5.5

When a body undergoes the deformation (b) of the previous example, show that there is no volume change.

In the reference configuration a line element PQ of length dL lies in the plane $X_1 = 0$, and is inclined at an angle θ to the X_3-axis, so that

$$dX_1 = 0, \quad dX_2 = dL \sin \theta, \quad dX_3 = dL \cos \theta$$

Show that the length dl of this element after the deformation is given by $dl = \{1+\kappa^2+\kappa \sin 2\theta\}^{\frac{1}{2}} dL$. Hence show that all such elements are extended provided $\kappa > 1$.

1.6   Non-homogeneous deformations

Deformations which are not of the form (1.5.1) are referred to as non-homogeneous deformations. We now discuss two such deformations which may be applied to either a solid or hollow circular cylinder. In each case we take our coordinate system such that the X_3-axis coincides with the axis of the cylinder and the base lies in the plane $X_3 = 0$.

(i) Simple torsion

Consider the deformation in which each cross-section remains in its original plane but is rotated through an angle τX_3 about the 3-axis, where τ is a constant called the twist per unit length (see Fig. 1.4). This deformation is referred to as *simple torsion*. Since each cross-section remains in its original plane,

$$x_3 = X_3 \qquad (1.6.1)$$

To find the remaining equations which specify this deformation, we consider the typical cross-section, shown in Fig. 1.5, which is at a distance X_3 from the base of the cylinder. If P is the particle

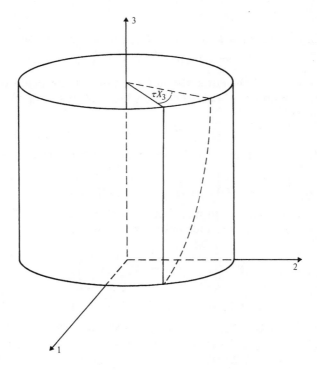

Fig. 1.4

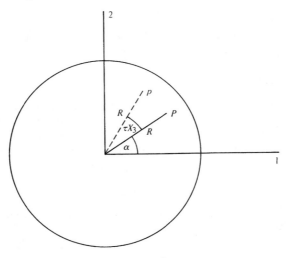

Fig. 1.5

whose initial coordinates are (X_1, X_2) then we may write

$$X_1 = R \cos \alpha, \qquad X_2 = R \sin \alpha \qquad (1.6.2)$$

where $R = (X_1^2 + X_2^2)^{\frac{1}{2}}$. After the deformation this particle occupies the point p with coordinates (x_1, x_2) where, from the figure it follows that

$$x_1 = R \cos(\tau X_3 + \alpha), \qquad x_2 = R \sin(\tau X_3 + \alpha) \qquad (1.6.3)$$

Expanding the sine and cosine functions and using (1.6.2) we obtain

$$x_1 = cX_1 - sX_2, \qquad x_2 = sX_1 + cX_2 \qquad (1.6.4)$$

where $c = \cos \tau X_3$, $s = \sin \tau X_3$. For the deformation specified by (1.6.1) and (1.6.4), the deformation gradient is given by

$$\mathbf{F} = \begin{pmatrix} c & -s & -\tau(sX_1 + cX_2) \\ s & c & \tau(cX_1 - sX_2) \\ 0 & 0 & 1 \end{pmatrix} \qquad (1.6.5)$$

so that $J = 1$, and the deformation is isochoric. Further,

$$\mathbf{B} = \begin{pmatrix} 1 + \tau^2(sX_1 + cX_2)^2 & -\tau^2(sX_1 + cX_2)(cX_1 - sX_2) & -\tau(sX_1 + cX_2) \\ -\tau^2(sX_1 + cX_2)(cX_1 - sX_2) & 1 + \tau^2(cX_1 - sX_2)^2 & \tau(cX_1 - sX_2) \\ -\tau(sX_1 + cX_2) & \tau(cX_1 - sX_2) & 1 \end{pmatrix}$$

$$(1.6.6)$$

Since $J = 1$, $\det \mathbf{B} = J^2 = 1$, and using (1.5.15) it follows that

$$\mathbf{B}^{-1} = \begin{pmatrix} 1 & 0 & \tau(sX_1 + cX_2) \\ 0 & 1 & -\tau(cX_1 - sX_2) \\ \tau(sX_1 + cX_2) & -\tau(cX_1 - sX_2) & 1 + \tau^2(X_1^2 + X_2^2) \end{pmatrix} \quad (1.6.7)$$

Using (1.4.28), (1.6.6), and (1.6.7), we obtain

$$I_1 = I_2 = 3 + \tau^2 R^2 \quad (1.6.8)$$

(ii) Torsion, extension, and inflation

Finally, we discuss the deformation which corresponds to simple extension along the axis of the cylinder, followed by simple torsion about its axis with twist τ per unit length. As a result of a uniform extension along the axis, the particle $\mathbf{X}$ is displaced to $\mathbf{X}'$, where

$$X_1' = \beta X_1, \quad X_2' = \beta X_2, \quad X_3' = \alpha X_3 \quad (1.6.9)$$

and here we take $\alpha > 0$, $\beta > 0$. If we now apply simple torsion to the extended cylinder, using (1.6.1) and (1.6.4), the final position $\mathbf{x}$ of the particle $\mathbf{X}$ is given by

$$\begin{aligned} x_1 &= X_1' \cos(\tau X_3') - X_2' \sin(\tau X_3'), \\ x_2 &= X_1' \sin(\tau X_3') + X_2' \cos(\tau X_3'), \quad x_3 = X_3' \end{aligned} \quad (1.6.10)$$

Combining (1.6.9) and (1.6.10), we obtain the deformation

$$\begin{aligned} x_1 &= \beta\{X_1 \cos(\alpha\tau X_3) - X_2 \sin(\alpha\tau X_3)\}, \\ x_2 &= \beta\{X_1 \sin(\alpha\tau X_3) + X_2 \cos(\alpha\tau X_3)\}, \quad x_3 = \alpha X_3 \end{aligned} \quad (1.6.11)$$

As a result of this deformation, the length of the cylinder increases or decreases depending on whether $\alpha > 1$ or $\alpha < 1$. Also, since

$$x_1^2 + x_2^2 = \beta^2(X_1^2 + X_2^2) \quad (1.6.12)$$

the radius of the cylinder increases if $\beta > 1$ and decreases if $\beta < 1$. The deformation is usually referred to as torsion, extension, and inflation. The deformation gradient is given by

$$\mathbf{F} = \begin{pmatrix} \beta c & -\beta s & -\alpha\tau\beta(sX_1 + cX_2) \\ \beta s & \beta c & \alpha\tau\beta(cX_1 - sX_2) \\ 0 & 0 & \alpha \end{pmatrix} \quad (1.6.13)$$

where $s = \sin \alpha\tau X_3$, $c = \cos \alpha\tau X_3$, so that $J = \alpha\beta^2$. If the material is incompressible, only isochoric deformations are possible, in which case

$$\beta = \alpha^{-\frac{1}{2}} \qquad (1.6.14)$$

Then

$B =$

$$\begin{pmatrix} \alpha^{-1} + \alpha\tau^2(sX_1 + cX_2)^2 & -\alpha\tau^2(sX_1 + cX_2)(cX_1 - sX_2) & -\alpha^{\frac{3}{2}}\tau(sX_1 + cX_2) \\ -\alpha\tau^2(cX_1 - sX_2)(sX_1 + cX_2) & \alpha^{-1} + \alpha\tau^2(cX_1 - sX_2)^2 & \alpha^{\frac{3}{2}}\tau(cX_1 - sX_2) \\ -\alpha^{\frac{3}{2}}\tau(sX_1 + cX_2) & \alpha^{\frac{3}{2}}\tau(cX_1 - sX_2) & \alpha^2 \end{pmatrix}$$

$$(1.6.15)$$

Since for the isochoric deformation $J = 1$, det $\mathbf{B} = 1$, and

$$\mathbf{B}^{-1} = \begin{pmatrix} \alpha & 0 & \alpha^{\frac{1}{2}}\tau(sX_1 + cX_2) \\ 0 & \alpha & -\alpha^{\frac{1}{2}}\tau(cX_1 - sX_2) \\ \alpha^{\frac{1}{2}}\tau(sX_1 + cX_2) & -\alpha^{\frac{1}{2}}\tau(cX_1 - sX_2) & \alpha^{-2}\{1 + \alpha^2\tau^2(X_1^2 + X_2^2)\} \end{pmatrix}$$

$$(1.6.16)$$

Also $I_3 = 1$, and from (1.6.15) and (1.6.16), using (1.4.28) it follows that

$$I_1 = \alpha^2 + 2\alpha^{-1} + \alpha\tau^2 R^2, \qquad I_2 = 2\alpha + \alpha^{-2} + \tau^2 R^2 \quad (1.6.17)$$

1.7 The displacement vector and infinitesimal strain tensor

Instead of specifying the current position **x** of the particle **X**, it is often convenient to consider its *displacement* **u** (see Fig. 1.6):

$$\mathbf{u} = \mathbf{x} - \mathbf{X} \qquad (1.7.1)$$

Then

$$\mathbf{x} = \mathbf{X} + \mathbf{u}$$

and

$$\mathbf{F} = \mathbf{I} + \mathbf{H} \qquad (1.7.2)$$

where **H** denotes the *displacement-gradient tensor* with components $(u_{K,L})$. The Cauchy–Green strain tensor **C** may now be written in the form

$$\mathbf{C} = \mathbf{F}^T\mathbf{F} = (\mathbf{I} + \mathbf{H}^T)(\mathbf{I} + \mathbf{H}) = \mathbf{I} + \mathbf{H} + \mathbf{H}^T + \mathbf{H}^T\mathbf{H} \qquad (1.7.3)$$

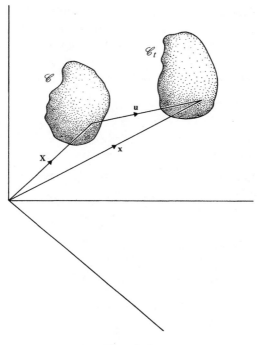

Fig. 1.6

or, in components,

$$C_{KL} = \delta_{KL} + u_{K,L} + u_{L,K} + u_{M,K} u_{M,L} \tag{1.7.4}$$

We have seen that $\mathbf{C}$ is a measure of the stretching part of the deformation. If $\mathbf{C} = \mathbf{I}$, then the particles are moving as a rigid body and so in general the components of $\mathbf{C} - \mathbf{I}$ may be thought of as measures of the change in shape. In some theories of continuum mechanics, particularly the infinitesimal theory of elasticity, $\mathbf{C} - \mathbf{I}$ is approximated by $\mathbf{H} + \mathbf{H}^T$; the product term $\mathbf{H}^T \mathbf{H}$ is neglected. In elasticity such an approximation appears to be satisfactory in a wide variety of applications in which the displacement gradients are small compared with the degree of accuracy required. The tensor

$$\mathbf{E} = \tfrac{1}{2}(\mathbf{H} + \mathbf{H}^T) \tag{1.7.5}$$

with components $E_{KL} = \tfrac{1}{2}(u_{K,L} + u_{L,K})$ is called the *infinitesimal strain tensor*.

For later use we derive here the form of the displacement field when the body is subjected to a rigid-body displacement for which $\mathbf{H}^{\mathrm{T}}\mathbf{H}$ may be neglected. When the body undergoes a rigid displacement, the final configuration may be obtained from the initial configuration by a rotation followed by a translation. In other words,

$$x_i = Q_{iK}X_K + c_i \qquad (1.7.6)$$

where $\mathbf{Q}$ is a constant proper orthogonal, or rotation, tensor and $\mathbf{c}$ is a constant vector. For this deformation,

$$\mathbf{F} = \mathbf{Q}, \quad \mathbf{U} = \mathbf{I}, \quad \mathbf{C} = \mathbf{I}, \quad \mathbf{R} = \mathbf{Q}, \quad \mathbf{H} = \mathbf{Q} - \mathbf{I} \qquad (1.7.7)$$

and

$$\mathbf{u} = \mathbf{x} - \mathbf{X} = (\mathbf{Q} - \mathbf{I})\mathbf{X} + \mathbf{c} \qquad (1.7.8)$$

Hence,

$$\mathbf{u} = \mathbf{H}\mathbf{X} + \mathbf{c} \qquad (1.7.9)$$

From (1.7.3) we see that, neglecting $\mathbf{H}^{\mathrm{T}}\mathbf{H}$,

$$\mathbf{H} + \mathbf{H}^{\mathrm{T}} = \mathbf{0} \qquad (1.7.10)$$

which means that $\mathbf{H}$ is skew-symmetric. Let $\boldsymbol{\omega} = (\omega_1, \omega_2, \omega_3)$ be the associated axial vector; that is,

$$\mathbf{H} = \begin{pmatrix} 0 & -\omega_3 & \omega_2 \\ \omega_3 & 0 & -\omega_1 \\ -\omega_2 & \omega_1 & 0 \end{pmatrix} \qquad (1.7.11)$$

Then from (1.7.9),

$$\mathbf{u} = \boldsymbol{\omega} \times \mathbf{X} + \mathbf{c} \qquad (1.7.12)$$

Example 1.7.1

Integrate the equations

$$E_{KL} = 0$$

to show that $\mathbf{u}$ must have the form (1.7.12).

Example 1.7.2

Show that

$$\mathbf{B} - \mathbf{I} = 2\mathbf{E} + \mathbf{H}\mathbf{H}^{\mathrm{T}}$$

1.8 Geometrical interpretation of the infinitesimal strains

Consider two line elements $d\mathbf{X}^{(1)}$ and $d\mathbf{X}^{(2)}$ at a point P in the reference configuration which have lengths dL_1, dL_2 and lie in the 1- and 2-directions. That is,

$$d\mathbf{X}^{(1)} = (dL_1, 0, 0), \qquad d\mathbf{X}^{(2)} = (0, dL_2, 0) \qquad (1.8.1)$$

These deform into vector line elements $d\mathbf{x}^{(1)}$ and $d\mathbf{x}^{(2)}$ at a point p in $\mathscr{C}_t$ and have lengths dl_1, dl_2 respectively (see Fig. 1.7), where, using (1.3.2),

$$dx_i^{(1)} = \frac{\partial x_i}{\partial X_1} dL_1, \qquad dx_i^{(2)} = \frac{\partial x_i}{\partial X_2} dL_2 \qquad (1.8.2)$$

In terms of the displacement $\mathbf{u}$, we may write

$$d\mathbf{x}^{(1)} = \left(1 + \frac{\partial u_1}{\partial X_1}, \frac{\partial u_2}{\partial X_1}, \frac{\partial u_3}{\partial X_1}\right) dL_1, \qquad d\mathbf{x}^{(2)} = \left(\frac{\partial u_1}{\partial X_2}, 1 + \frac{\partial u_2}{\partial X_2}, \frac{\partial u_3}{\partial X_2}\right) dL_2$$

$$(1.8.3)$$

Thus

$$(dl_1)^2 = d\mathbf{x}^{(1)} \cdot d\mathbf{x}^{(1)} = \left(1 + 2\frac{\partial u_1}{\partial X_1}\right)(dL_1)^2$$

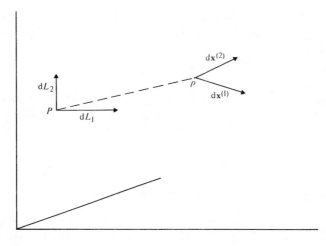

Fig. 1.7

to the first order in the displacement gradients. Hence

$$dl_1 = \left(1 + 2\frac{\partial u_1}{\partial X_1}\right)^{\frac{1}{2}} dL_1 = (1 + E_{11})\, dL_1 \qquad (1.8.4)$$

to the same order of approximation, using the definition (1.7.5). This may be written in the form

$$\frac{dl_1 - dL_1}{dL_1} = E_{11} \qquad (1.8.5)$$

The component E_{11} of the infinitesimal strain tensor therefore measures the proportional extension of a line element which is initially in the 1-direction. The components E_{22} and E_{33} have similar interpretations. To understand the meaning of E_{12}, we form the scalar product

$$
\begin{aligned}
dx^{(1)} \cdot dx^{(2)} &= \left(1 + \frac{\partial u_1}{\partial X_1}, \frac{\partial u_2}{\partial X_1}, \frac{\partial u_3}{\partial X_1}\right) \cdot \left(\frac{\partial u_1}{\partial X_2}, 1 + \frac{\partial u_2}{\partial X_2}, \frac{\partial u_3}{\partial X_2}\right) dL_1\, dL_2 \\
&= \left(\frac{\partial u_1}{\partial X_2} + \frac{\partial u_2}{\partial X_1}\right) dL_1\, dL_2 \qquad (1.8.6)
\end{aligned}
$$

to the first order in the displacement gradients. Or,

$$dl_1\, dl_2 \cos\theta = 2E_{12}\, dL_1\, dL_2 \qquad (1.8.7)$$

where θ is the angle between $dx^{(1)}$ and $dx^{(2)}$. In view of (1.8.4), and a similar expression for dl_2, equation (1.8.7) may be written, to first order, as

$$E_{12} = \tfrac{1}{2}\cos\theta \qquad (1.8.8)$$

In other words, E_{12} is half the cosine of the angle between the directions into which elements along the 1- and 2-directions in $\mathscr{C}$ are deformed. Similar interpretations can be given to E_{23} and E_{31}.

1.9   The continuity equation

Some of the physical principles which apply to the behaviour of all materials can be expressed in the form of *balance laws*. Such laws normally equate the total rate of increase of a certain physical quantity in any given part of the body to its rate of supply, and this rate of supply usually arises from a surface flux together with a volume distribution. We state these laws for an arbitrary part $\mathscr{P}$ of the body, so that where volume and surface integrals of quantities in the spatial description are involved, they are calculated for given sets of particles, and so the regions of integration are moving with the material. We refer to these regions as *material volumes* and *material surfaces*. We show how these balance laws can be reduced to point form, that is, to differential equations which hold at each point of the body.

The simplest of these laws concerns the balance of mass. Consider any material volume V in $\mathscr{C}_t$, bounded by a surface S. We emphasise that in stating the law we are considering a given set of particles $\mathscr{P}$, and V is the region they happen to occupy at time t. We assume that no mechanism exists for the creation of mass throughout the region of the body, and since a material surface moves with the particles of the body, there is no flux of mass across S. Hence, the rate of increase of the total mass of $\mathscr{P}$ is zero.

Let ρ denote the density of the body in the current configuration $\mathscr{C}_t$, and ρ_0 the density in $\mathscr{C}_0$ which we take to be the reference configuration $\mathscr{C}$. Then the total mass of $\mathscr{P}$ is

$$\int_V \rho \, dV \qquad (1.9.1)$$

Since the rate of increase of this mass is zero throughout the motion, the total mass of $\mathscr{P}$ in $\mathscr{C}_t$ must be equal to its mass in $\mathscr{C}_0$. Hence,

$$\int_V \rho \, dV = \int_{V_0} \rho_0 \, dV_0 \qquad (1.9.2)$$

where V_0 is the region of $\mathscr{C}_0$ occupied by the particles of $\mathscr{P}$. Equation (1.9.2) is the integral statement of the law concerning balance of mass. To reduce it to point form, we first observe that the volume integral on the left-hand side may be expressed as an integral over the corresponding region V_0 of $\mathscr{C}_0$ using the Jacobian J, and relation (1.3.5), that is,

$$\int_V \rho \, dV = \int_{V_0} \rho J \, dV_0$$

Hence, (1.9.2) becomes

$$\int_{V_0} (\rho J - \rho_0)\, dV_0 = 0 \tag{1.9.3}$$

We assume that the integrand in (1.9.3) is continuous, and so if this integrand is positive at any point it must be positive in a neighbourhood about that point. Since $\mathcal{P}$ was chosen as an arbitrary part of the body, (1.9.3) must hold for all regions V_0 of $\mathcal{C}_0$ and so, in particular, it holds for the neighbourhood in which the integrand has been presumed to be positive, and this is clearly impossible. Thus our assumption that there exists a point at which the integrand is positive must be false. A similar argument shows that the integrand cannot be negative. Hence we must have

$$\rho J - \rho_0 = 0 \tag{1.9.4}$$

Now ρ_0, being the density in $\mathcal{C}_0$, depends only on $\mathbf{X}$ and so

$$\frac{D}{Dt}(\rho J) = 0 \tag{1.9.5}$$

Using (1.3.6) we find

$$\frac{D\rho}{Dt} + \rho\, \text{div } \mathbf{v} = 0 \tag{1.9.6}$$

or using (1.2.4) this can be written

$$\frac{\partial \rho}{\partial t} + \text{div }(\rho \mathbf{v}) = 0 \tag{1.9.7}$$

Equations (1.9.4), (1.9.6), and (1.9.7) are equivalent point forms of the balance law expressing conservation of mass. They are usually referred to as *continuity equations*.

Example 1.9.1

Show that for any scalar, vector, or tensor field ψ, and any material volume V,

$$\frac{d}{dt}\int_{V} \rho\psi\, dV = \int_{V} \rho\frac{D\psi}{Dt}\, dV \tag{1.9.8}$$

We may again transform the volume integral on the left-hand side of (1.9.8) into a volume integral over V_0 using the Jacobian

J. Thus

$$\int_V \rho\psi \, \mathrm{d}V = \int_{V_0} \rho J\psi \, \mathrm{d}V_0 \qquad (1.9.9)$$

Since the integrals are taken over a given set of particles, V_0 is fixed in time, so that

$$\frac{\mathrm{d}}{\mathrm{d}t}\int_V \rho\psi \, \mathrm{d}V = \int_{V_0} \left\{ \psi\frac{\mathrm{D}}{\mathrm{D}t}(\rho J) + \rho J\frac{\mathrm{D}\psi}{\mathrm{D}t} \right\} \mathrm{d}V_0 \qquad (1.9.10)$$

The material time derivatives appear on the right-hand side of (1.9.10) since the integrand is in the material description. Using (1.9.5) and transforming back to V we find

$$\frac{\mathrm{d}}{\mathrm{d}t}\int_V \rho\psi \, \mathrm{d}V = \int_V \rho\frac{\mathrm{D}\psi}{\mathrm{D}t} \, \mathrm{d}V$$

Example 1.9.2

For a material volume V bounded by a surface S with outward unit normal $\mathbf{n}$, show that

$$\frac{\mathrm{d}}{\mathrm{d}t}\int_V \psi \, \mathrm{d}V = \int_V \frac{\partial\psi}{\partial t} \, \mathrm{d}V + \int_S \psi\mathbf{v}\cdot\mathbf{n} \, \mathrm{d}S \qquad (1.9.11)$$

The terms appearing in (1.9.11) have an interesting interpretation when ψ is, for example, the density ρ. Let v be the region fixed in space which instantaneously coincides with the material volume V at time t. Then

$$\frac{\mathrm{d}}{\mathrm{d}t}\int_v \rho \, \mathrm{d}v = \int_v \frac{\partial\rho}{\partial t} \, \mathrm{d}v = \int_V \frac{\partial\rho}{\partial t} \, \mathrm{d}V$$

$$= \frac{\mathrm{d}}{\mathrm{d}t}\int_V \rho \, \mathrm{d}V - \int_S \rho\mathbf{v}\cdot\mathbf{n} \, \mathrm{d}S$$

$$= \frac{\mathrm{d}}{\mathrm{d}t}\int_V \rho \, \mathrm{d}V - \int_s \rho\mathbf{v}\cdot\mathbf{n} \, \mathrm{d}s$$

where s is the surface, fixed in space, instantaneously coinciding with S at time t. This means that the rate of increase of mass in the fixed region v is equal to the rate of increase of mass in the material region V less the rate of mass flow out of s.

Example 1.9.3

A continuous motion of the body is given by

$$x_1 = X_1 + \alpha t X_3, \quad x_2 = X_2 + \alpha t X_3, \quad x_3 = X_3 - \alpha t (X_1 + X_2)$$

where α is a constant. If ρ_0 is the uniform density in the reference configuration, show that the density in the current configuration is everywhere equal to $\rho_0/(1 + 2\alpha^2 t^2)$.

Example 1.9.4

The components of the velocity field in a liquid of constant density ρ are given, in the spatial description, by

$$v_1 = -\alpha x_2/(x_1^2 + x_2^2), \quad v_2 = \alpha x_1/(x_1^2 + x_2^2), \quad v_3 = 0$$

where $x_1^2 + x_2^2 \neq 0$ and α is a constant. Show that this velocity field satisfies the equation of continuity.

1.10   The stress vector and body force

The concept of force is a familiar one in particle and rigid-body dynamics. In those subjects a force usually acts at a point (gravitational forces on rigid bodies may be regarded as forces acting at the centre of gravity). In continuum mechanics we are additionally concerned with the interaction between neighbouring portions of the interiors of deformable bodies. In reality such interaction consists of complex interatomic forces, but we make the simplifying assumption that the effect of all such forces across any given surface may be adequately represented by a single vector field defined over the surface. We also assume that the effect of external forces such as gravity may be represented by another vector field defined throughout the region occupied by the body.

Let V again denote the region occupied at time t by an arbitrary part $\mathscr{P}$ of the body, and let S denote the closed surface bounding V. As usual, $\mathbf{n}$ is the outward-drawn unit normal to S and we postulate the existence of a vector field $\mathbf{t}(\mathbf{x}, \mathbf{n})$ defined over S, and a vector field $\mathbf{b}(\mathbf{x})$ defined over V, such that the total force on $\mathscr{P}$ is

$$\int_S \mathbf{t}(\mathbf{x}, \mathbf{n}) \, dS + \int_V \rho \mathbf{b}(\mathbf{x}) \, dV \tag{1.10.1}$$

and the total torque on $\mathscr{P}$ about the origin is

$$\int_S \mathbf{x} \times \mathbf{t}(\mathbf{x}, \mathbf{n}) \, dS + \int_V \rho \mathbf{x} \times \mathbf{b}(\mathbf{x}) \, dV \qquad (1.10.2)$$

The vector $\mathbf{t}(\mathbf{x}, \mathbf{n})$ is known as the *stress vector;* the indicated dependence on $\mathbf{x}$ and $\mathbf{n}$ is intended to convey that $\mathbf{t}$ varies with position and also with the orientation of the surface with which it is associated. At points where S is interior to the body, $\mathbf{t}(\mathbf{x}, \mathbf{n})$ represents the force per unit area on $\mathscr{P}$ exerted by the material outside $\mathscr{P}$. At points where S coincides with the surface of the body, if any, $\mathbf{t}(\mathbf{x}, \mathbf{n})$ represents the force per unit area applied to the surface of the body by an external agency, and we refer to this as *surface traction.* In place of $\mathbf{t}(\mathbf{x}, \mathbf{n})$ we often write $\mathbf{t}$ when the meaning is clear, or sometimes we write $\mathbf{t}(\mathbf{n})$ when we wish to emphasise the dependence on $\mathbf{n}$ at a particular point $\mathbf{x}$. The vector field $\mathbf{b}(\mathbf{x})$ is known as the *body force*, and represents the distributed force per unit mass acting throughout $\mathscr{P}$ due to an external agency, usually gravitation. We are considering here the body force acting at an instant of time t, but in general $\mathbf{b}(\mathbf{x})$ may vary with time.

1.11 Principles of linear and angular momentum. The stress tensor

When dealing with systems of particles, we can deduce from Newton's laws of motion, and certain other assumptions, that the resultant of the external forces acting on the system is equal to the total rate of change of the linear momentum of the system. By taking moments about a fixed point, we can also show that the resultant moment of the external forces is equal to the total rate of change of moment of momentum.

Here we define the linear and angular momentum density for a continuum and introduce balance laws for these quantities. In formulating these laws we are motivated by the corresponding results for systems of particles, but in continuum mechanics they are to be understood as fundamental principles. The justification for them in no way depends on particle mechanics, but lies entirely in the apparent usefulness of the theories which are founded on them.

We define the linear momentum density to be $\rho\mathbf{v}$ per unit volume, so that the total linear momentum of the part $\mathscr{P}$ of the body is

$$\int_V \rho\mathbf{v}\,dV \qquad (1.11.1)$$

We now take as our principle of linear momentum the assertion that the rate of change of the linear momentum of $\mathscr{P}$ is equal to the resultant force on $\mathscr{P}$. That is,

$$\frac{d}{dt}\int_V \rho\mathbf{v}\,dV = \int_S \mathbf{t}(\mathbf{x},\mathbf{n})\,dS + \int_V \rho\mathbf{b}(\mathbf{x})\,dV \qquad (1.11.2)$$

We emphasise that the integrals here are taken over a given set of particles $\mathscr{P}$, so that in calculating the time derivative on the left-hand side of (1.11.2) the variation of V with the motion of $\mathscr{P}$ must be taken into account. Using (1.9.8), (1.11.2) may be written

$$\int_V \rho\frac{D\mathbf{v}}{Dt}\,dV = \int_S \mathbf{t}(\mathbf{x},\mathbf{n})\,dS + \int_V \rho\mathbf{b}(\mathbf{x})\,dV \qquad (1.11.3)$$

Suppose now that the region V (Fig. 1.8) is cut into two parts V_1 and V_2 by a surface σ. The region V_1 is then bounded by σ

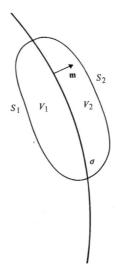

Fig. 1.8

and a part S_1 of S; similarly, V_2 is bounded by σ and a part S_2 of S. As (1.11.3) holds for any part of the body, it must hold for regions V_1 and V_2 separately. Using **m** to denote the unit normal to σ *out* of V_1 we have therefore

$$\int_{V_1} \rho \frac{D\mathbf{v}}{Dt} dV = \int_{S_1} \mathbf{t}(\mathbf{x}, \mathbf{n}) dS + \int_{\sigma} \mathbf{t}(\mathbf{x}, \mathbf{m}) dS + \int_{V_1} \rho\mathbf{b}(\mathbf{x}) dV \quad (1.11.4)$$

and

$$\int_{V_2} \rho \frac{D\mathbf{v}}{Dt} dV = \int_{S_2} \mathbf{t}(\mathbf{x}, \mathbf{n}) dS + \int_{\sigma} \mathbf{t}(\mathbf{x}, -\mathbf{m}) dS$$
$$+ \int_{V_2} \rho\mathbf{b}(\mathbf{x}) dV \quad (1.11.5)$$

Since $S = S_1 + S_2$ and $V = V_1 + V_2$, subtracting both sides of (1.11.4) and (1.11.5) from (1.11.3) we have

$$\int_{\sigma} \{\mathbf{t}(\mathbf{x}, \mathbf{m}) + \mathbf{t}(\mathbf{x}, -\mathbf{m})\} dS = \mathbf{0} \quad (1.11.6)$$

By the same argument, a similar result can be shown to hold for all subregions of σ. We therefore conclude that

$$\mathbf{t}(\mathbf{x}, \mathbf{m}) = -\mathbf{t}(\mathbf{x}, -\mathbf{m}) \quad (1.11.7)$$

As the choice of σ is arbitrary, (1.11.7) must hold for all surfaces in the continuum. It shows us that, at a given point, the stress vector acting on one side of a surface balances that on the other.

Example 1.11.1

Consider a tetrahedron, three faces of which are parallel to the coordinate planes and meet at an arbitrary point P of the body (see Fig. 1.9). The fourth face has area A and unit normal **n** in an arbitrary direction. Let the area of the face with unit normal $\mathbf{e}_i$ be A_i. Show that the angle between the face with normal $\mathbf{e}_i$ and the fourth face is $\cos^{-1} n_i$. Hence show that $A_i = n_i A$. Prove also that the volume of the tetrahedron is $\frac{1}{3}hA$, where h is the distance of P from the fourth face.

We now apply (1.11.3) to the tetrahedron described in Example 1.11.1 and proceed to the limit as $h \to 0$ keeping **n** fixed. We assume that $|\rho\, D\mathbf{v}/Dt|$ and $|\rho\mathbf{b}|$ are bounded and have maximum

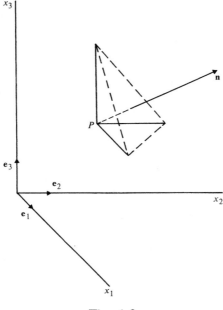

Fig. 1.9

values a, b respectively. Then

$$\left| \int_V \rho \frac{\mathbf{Dv}}{\mathbf{D}t} \, \mathrm{d}V \right| \leqslant \tfrac{1}{3} hAa$$

and

$$\left| \int_V \rho \mathbf{b} \, \mathrm{d}V \right| \leqslant \tfrac{1}{3} hAb$$

So that as $h \to 0$

$$\frac{1}{A} \int_V \rho \frac{\mathbf{Dv}}{\mathbf{D}t} \, \mathrm{d}V - \frac{1}{A} \int_V \rho \mathbf{b} \, \mathrm{d}V \to \mathbf{0}$$

Hence, dividing each term in (1.11.3) by A, we find that

$$\lim_{h \to 0} \frac{1}{A} \int_S \mathbf{t}(\mathbf{x}, \mathbf{n}) \, \mathrm{d}S = \mathbf{0}$$

where the surface integral is taken over the four faces of the tetrahedron. In the limit as $h \to 0$, the areas of the four faces tend to zero, and the surface integral over a face approaches the value of the integrand at P multiplied by the area of the face. Hence,

$$\mathbf{t}(\mathbf{x}, \mathbf{n}) + \mathbf{t}(\mathbf{x}, -\mathbf{e}_1)\frac{A_1}{A} + \mathbf{t}(\mathbf{x}, -\mathbf{e}_2)\frac{A_2}{A} + \mathbf{t}(\mathbf{x}, -\mathbf{e}_3)\frac{A_3}{A} = \mathbf{0} \quad (1.11.8)$$

Using (1.11.7) and the result $A_i = n_i A$ of Example 1.11.1, (1.11.8) becomes

$$\mathbf{t}(\mathbf{x}, \mathbf{n}) = n_1 \mathbf{t}(\mathbf{x}, \mathbf{e}_1) + n_2 \mathbf{t}(\mathbf{x}, \mathbf{e}_2) + n_3 \mathbf{t}(\mathbf{x}, \mathbf{e}_3) \quad (1.11.9)$$

If we write

$$\mathbf{t}(\mathbf{x}, \mathbf{e}_1) = T_{11}\mathbf{e}_1 + T_{12}\mathbf{e}_2 + T_{13}\mathbf{e}_3$$
$$\mathbf{t}(\mathbf{x}, \mathbf{e}_2) = T_{21}\mathbf{e}_1 + T_{22}\mathbf{e}_2 + T_{23}\mathbf{e}_3 \quad (1.11.10)$$
$$\mathbf{t}(\mathbf{x}, \mathbf{e}_3) = T_{31}\mathbf{e}_1 + T_{32}\mathbf{e}_2 + T_{33}\mathbf{e}_3$$

then (1.11.9) becomes

$$t_i(\mathbf{x}, \mathbf{n}) = T_{ji}n_j \quad (1.11.11)$$

The quantities T_{ij} form the components of a second-order tensor **T** known as the *Cauchy stress tensor*. From (1.11.9) (or (1.11.11)) we see that the dependence of $\mathbf{t}(\mathbf{x}, \mathbf{n})$ upon the unit normal **n** is linear and when the stress components T_{ij} are known, as well as the normal to the surface, the components of the stress vector can be calculated.

From (1.11.10) we see that T_{ij} denotes the component in the j-direction of the force per unit area acting on an element of surface in the deformed configuration which has normal in the i-direction. For example, T_{13} is the component in the 3-direction of the force per unit area on an element of surface with normal in the 1-direction. Figure 1.10 shows the stress components acting on the faces $x_1 = $ constant of a rectangular parallelepiped whose faces are parallel to the coordinate planes. In this figure we assume that T_{11}, T_{12}, and T_{13} are positive and we have used (1.11.7).

Example 1.11.2

Let $\mathbf{e}_i$ and $\mathbf{e}_i'$ be the base vectors of two Cartesian coordinate systems $Ox_1x_2x_3$ and $Ox_1'x_2'x_3'$. Then the coordinate transformation is given by

$$x_i' = l_{ij}x_j$$

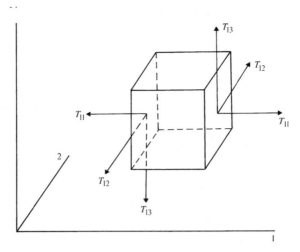

Fig. 1.10

where $l_{ij} = \mathbf{e}'_i \cdot \mathbf{e}_j$. Show that the stress components T_{ij}, T'_{ij} associated with the two coordinate systems are related by

$$T'_{ij} = l_{ip}l_{jq}T_{pq} \qquad (1.11.12)$$

This shows that T_{ij} are the components of a second-order tensor.

[*Hint:* Use the definition

$$T'_{ij} = \mathbf{t}(\mathbf{e}'_i) \cdot \mathbf{e}'_j$$

and the relation $\mathbf{e}'_i = l_{ij}\mathbf{e}_j$, together with (1.11.11).]

Example 1.11.3

A certain coordinate transformation is given by

$$\begin{pmatrix} x'_1 \\ x'_2 \\ x'_3 \end{pmatrix} = \begin{pmatrix} \cos\theta & \sin\theta & 0 \\ -\sin\theta & \cos\theta & 0 \\ 0 & 0 & 1 \end{pmatrix} \begin{pmatrix} x_1 \\ x_2 \\ x_3 \end{pmatrix}$$

Show that this transformation represents a rotation through an angle θ about Ox_3. If the components of the stress tensor **T** relative to the axes $Ox_1x_2x_3$ are

$$\begin{pmatrix} 1 & 0 & 1 \\ 0 & 1 & 1 \\ 1 & 1 & 1 \end{pmatrix}$$

in some specified units, find the components of the stress tensor relative to $Ox_1'x_2'x_3'$.

The stress vector $\mathbf{t}$ may be resolved into a component ν in the direction of $\mathbf{n}$ and a component σ in the direction of $\mathbf{s}$, where $\mathbf{s}$ is a unit vector perpendicular to $\mathbf{n}$ so that

$$\nu = \mathbf{t} \cdot \mathbf{n}, \qquad \sigma = \mathbf{t} \cdot \mathbf{s} \qquad (1.11.13)$$

The component σ therefore represents the tangential component of the stress vector acting on the element of surface at $\mathbf{x}$ with unit normal $\mathbf{n}$, and ν is the normal component; σ is sometimes referred to as a shear stress and ν as a normal stress. In particular, from (1.11.10) we see that T_{ij} $(i = j)$ and T_{ij} $(i \neq j)$ represent normal and shear stresses respectively on the coordinate planes.

Example 1.11.4

The stress tensor at a certain point in a body has components

$$\begin{pmatrix} 1 & 1 & 0 \\ 1 & -1 & 0 \\ 0 & 0 & 1 \end{pmatrix}$$

in some specified units. Find the stress vector acting on an element of surface with normal in the direction of $(1, 1, 2)$. Show that the normal stress has magnitude 1. Show also that the shear stress has magnitude $1/\sqrt{3}$ and acts in the direction $(1, -1, 0)$.

The unit normal $\mathbf{n}$ to the surface element has components $(1/\sqrt{6})(1, 1, 2)$ and so, using (1.11.11), the stress vector $\mathbf{t}$ has components

$$(1/\sqrt{6}) \begin{pmatrix} 1 & 1 & 0 \\ 1 & -1 & 0 \\ 0 & 0 & 1 \end{pmatrix} \begin{pmatrix} 1 \\ 1 \\ 2 \end{pmatrix} = \frac{1}{\sqrt{6}} \begin{pmatrix} 2 \\ 0 \\ 2 \end{pmatrix}$$

Hence, the normal stress is

$$\nu = \mathbf{t} \cdot \mathbf{n} = \tfrac{1}{6} \cdot 6 = 1$$

Now

$$\mathbf{t} = \nu \mathbf{n} + \mathbf{s}$$

Thus

$$\mathbf{s} = \mathbf{t} - \nu\mathbf{n} = (1/\sqrt{6})(1, -1, 0)$$

and so $\mathbf{s}$ has magnitude $\sigma = \sqrt{2}/\sqrt{6} = 1/\sqrt{3}$ and acts in the direction $(1, -1, 0)$.

Substituting (1.11.11) into (1.11.3) we find

$$\int_S T_{ji} n_j \, \mathrm{d}S + \int_V \rho b_i \, \mathrm{d}V = \int_V \rho \frac{\mathrm{D}v_i}{\mathrm{D}t} \mathrm{d}V$$

and, using the divergence theorem this becomes

$$\int_V T_{ji,j} \, \mathrm{d}V + \int_V \rho b_i \, \mathrm{d}V = \int_V \rho \frac{\mathrm{D}v_i}{\mathrm{D}t} \mathrm{d}V$$

Since this holds for all regions V of the body, and assuming that the integrand is continuous, we obtain finally

$$T_{ji,j} + \rho b_i = \rho \frac{\mathrm{D}v_i}{\mathrm{D}t} \tag{1.11.14}$$

The density of angular momentum, or moment of momentum, about the origin is defined to be $\mathbf{x} \times \rho\mathbf{v}$ per unit volume so that the total angular momentum of $\mathscr{P}$ is

$$\int_V (\mathbf{x} \times \rho\mathbf{v}) \, \mathrm{d}V$$

The principle of angular momentum asserts that the total torque on $\mathscr{P}$ about a fixed point which we take to be the origin, is equal to the rate of change of its angular momentum about this point. That is,

$$\int_S \mathbf{x} \times \mathbf{t} \, \mathrm{d}S + \int_V \mathbf{x} \times \rho\mathbf{b} \, \mathrm{d}V = \frac{\mathrm{d}}{\mathrm{d}t} \int_V \mathbf{x} \times \rho\mathbf{v} \, \mathrm{d}V \tag{1.11.15}$$

In component form this may be written

$$\int_S \varepsilon_{ijk} x_j t_k \, \mathrm{d}S + \int_V \varepsilon_{ijk} x_j \rho b_k \, \mathrm{d}V = \frac{\mathrm{d}}{\mathrm{d}t} \int_V \varepsilon_{ijk} x_j \rho v_k \, \mathrm{d}V \tag{1.11.16}$$

(see Example 1.3.3 for the definition of ε_{ijk}). Using (1.9.8), (1.11.11), and the divergence theorem, (1.11.16) may be written

$$\int_V \varepsilon_{ijk} \left\{ \frac{\partial}{\partial x_r} (x_j T_{rk}) + \rho x_j b_k - \rho \frac{\mathrm{D}}{\mathrm{D}t} (x_j v_k) \right\} \mathrm{d}V = 0 \tag{1.11.17}$$

Now

$$\varepsilon_{ijk} \frac{\partial}{\partial x_r} (x_j T_{rk}) = \varepsilon_{ijk} (\delta_{jr} T_{rk} + x_j T_{rk,r})$$

$$= \varepsilon_{ijk} (T_{jk} + x_j T_{rk,r})$$

and

$$\varepsilon_{ijk} \frac{D}{Dt} (x_j v_k) = \varepsilon_{ijk} \left(v_j v_k + x_j \frac{D v_k}{Dt} \right)$$

$$= \varepsilon_{ijk} x_j \frac{D v_k}{Dt}$$

So, using (1.11.14), equation (1.11.17) reduces to

$$\int_V \varepsilon_{ijk} T_{jk} \, dV = 0$$

This is again a relation which holds for all regions V. Hence,

$$\varepsilon_{ijk} T_{jk} = 0$$

Putting $i = 1, 2, 3$ in turn we find $T_{23} - T_{32} = 0$, $T_{31} - T_{13} = 0$, $T_{12} - T_{21} = 0$. That is,

$$T_{ij} = T_{ji} \tag{1.11.18}$$

In other words, the stress tensor is symmetric.

Example 1.11.5

Show that the component of $\mathbf{t}(\mathbf{x}, \mathbf{n})$ in the direction of $\mathbf{n}^*$ is equal to the component of $\mathbf{t}(\mathbf{x}, \mathbf{n}^*)$ in the direction of $\mathbf{n}$.

We conclude this section with a discussion of some important special states of stress.

(i) Purely normal stress

When the stress tensor takes the form

$$\mathbf{T} = -p\mathbf{I}, \qquad T_{ij} = -p\delta_{ij} \tag{1.11.19}$$

where p is a constant, the stress vector acting on a surface element with arbitrary unit normal $\mathbf{n}$ is

$$\mathbf{t}(\mathbf{n}) = -p\mathbf{n} \tag{1.11.20}$$

Such a stress system is therefore said to be *purely normal*; if $p > 0$, the phrase *hydrostatic pressure* is sometimes used. If the material is in equilibrium and body force is absent, the equation of motion (1.11.14) is satisfied identically.

(ii) Uniaxial tension and compression

When T_{11} equals a constant T and is the only non-vanishing component of the stress tensor, so that

$$\mathbf{T} = \begin{pmatrix} T & 0 & 0 \\ 0 & 0 & 0 \\ 0 & 0 & 0 \end{pmatrix} \tag{1.11.21}$$

this state of stress is called a *uniaxial tension* T in the 1-direction if $T > 0$, and a *uniaxial compression* T in the 1-direction if $T < 0$. A uniform normal stress T acts on every plane with normal in the 1-direction; and all planes with normals perpendicular to the 1-direction are free of traction.

A uniaxial tension or compression in the 2- or 3-direction may be defined similarly. By rotation of axes and use of (1.11.12), the stress components corresponding to a uniaxial tension T in an arbitrary direction $\mathbf{n}$ may be calculated.

Example 1.11.6

Show that the stress components associated with a uniaxial tension T in the direction of the unit vector $\mathbf{n}$ are

$$T \begin{pmatrix} n_1^2 & n_1 n_2 & n_1 n_3 \\ n_2 n_1 & n_2^2 & n_2 n_3 \\ n_3 n_1 & n_3 n_2 & n_3^2 \end{pmatrix} \tag{1.11.22}$$

Example 1.11.7

If $\mathbf{m}$, $\mathbf{n}$ are unit vectors, and the material is subjected to a uniaxial tension T in the $\mathbf{n}$-direction, show that

$$\mathbf{t}(\mathbf{m}) = T(\mathbf{m} \cdot \mathbf{n})\mathbf{n} \tag{1.11.23}$$

(iii) Pure shear

When T_{12} equals a constant S, and is the only non-vanishing stress component, so that

$$\mathbf{T} = \begin{pmatrix} 0 & S & 0 \\ S & 0 & 0 \\ 0 & 0 & 0 \end{pmatrix} \tag{1.11.24}$$

this state of stress is called a *pure shear* relative to the 1- and 2-directions. A surface with normal in the 1-direction experiences a shear stress in the 2-direction. A surface with normal in the 2-direction experiences a shear stress in the 1-direction. A surface with normal in the 3-direction is free of traction. Pure shears relative to the 2- and 3-directions and the 3- and 1-directions may be defined similarly.

Example 1.11.8

Show that the stress components associated with a pure shear S relative to the directions defined by orthogonal unit vectors **n**, **m** are given by

$$S\begin{pmatrix} 2n_1m_1 & n_1m_2+n_2m_1 & n_1m_3+n_3m_1 \\ n_1m_2+n_2m_1 & 2n_2m_2 & n_2m_3+n_3m_2 \\ n_1m_3+n_3m_1 & n_2m_3+n_3m_2 & 2n_3m_3 \end{pmatrix} \quad (1.11.25)$$

Example 1.11.9

If **m**, **n** are orthogonal unit vectors and **p** is a third unit vector, and the material is subjected to a pure shear S relative to **m**, **n**, show that

$$\mathbf{t}(\mathbf{p}) = S\{(\mathbf{m} \cdot \mathbf{p})\mathbf{n} + (\mathbf{n} \cdot \mathbf{p})\mathbf{m}\} \quad (1.11.26)$$

Example 1.11.10

Show that a pure shear S relative to the 1- and 2-directions is equivalent to the superposition of a uniaxial tension S and a uniaxial compression S both in the $(1, 2)$-plane and making angles $\frac{1}{4}\pi$ and $\frac{3}{4}\pi$, respectively, with the 1-direction.

1.12 Principal stresses. Principal axes of stress. Stress invariants

Consider an element of surface dS at a point P of $\mathscr{C}_t$. Let **n** be the unit normal to one of the sides of dS. Then the force per unit area acting on that side of dS is $\mathbf{t}(\mathbf{x}, \mathbf{n})$. This vector is in the direction of **n** only for those directions **n** for which

$$\mathbf{t}(\mathbf{x}, \mathbf{n}) = \lambda \mathbf{n} \quad (1.12.1)$$

for some scalar λ; or, using (1.11.11) and (1.11.18),

$$(T_{ij} - \lambda \delta_{ij})n_j = 0 \tag{1.12.2}$$

The only directions **n** for which this is possible lie along the principal axes of **T** which are known as the *principal axes of stress*. The associated values of λ are the principal values of **T** and these are known as the *principal stresses*. They are the real roots of the characteristic equation

$$\det(T_{ij} - \lambda \delta_{ij}) = 0 \tag{1.12.3}$$

The principal stress invariants are defined from the characteristic equation in the same manner as for strains (see (1.4.23)), and are denoted by

$$\begin{aligned}
J_1 &= \operatorname{tr} \mathbf{T} = T_{ii} \\
J_2 &= \tfrac{1}{2}(\operatorname{tr} \mathbf{T})^2 - \tfrac{1}{2}\operatorname{tr} \mathbf{T}^2 = \tfrac{1}{2}(T_{ii}T_{jj} - T_{ij}T_{ij}) \\
J_3 &= \det \mathbf{T}
\end{aligned} \tag{1.12.4}$$

Example 1.12.1

  (i) Using Example 1.11.5, show that, if the principal stresses are distinct, there exists a set of three mutually perpendicular principal axes.
 (ii) If two principal stresses coincide, but are distinct from the third, show that there is a principal axis associated with the third principal stress, and all axes perpendicular to this axis are also principal axes.
(iii) If all three principal stresses coincide, show that all axes are principal axes.

(For further discussion of principal stresses and principal axes of stress, see Spencer (1980) Section 5.6.)

1.13   The energy-balance equation

Taking the scalar product of the equation of motion (1.11.14) with **v**, we have

$$\begin{aligned}
\rho v_i \frac{\mathrm{D}v_i}{\mathrm{D}t} &= v_i T_{ji,j} + \rho b_i v_i \\
&= (v_i T_{ji})_{,j} - v_{i,j} T_{ji} + \rho b_i v_i
\end{aligned} \tag{1.13.1}$$

Now suppose that we split the components $v_{i,j}$ of the velocity-gradient tensor into symmetric and skew-symmetric parts by writing

$$v_{i,j} = D_{ij} + W_{ij} \tag{1.13.2}$$

where

$$D_{ij} = \tfrac{1}{2}(v_{i,j} + v_{j,i}), \qquad W_{ij} = \tfrac{1}{2}(v_{i,j} - v_{j,i}) \tag{1.13.3}$$

Then D_{ij} and W_{ij} are the components of the *rate-of-deformation tensor* **D** and the *spin tensor* **W**, and in view of (1.11.18), equation (1.13.1) can be written

$$\rho v_i \frac{Dv_i}{Dt} + T_{ij}D_{ij} = (v_i T_{ij})_{,j} + \rho b_i v_i \tag{1.13.4}$$

Integrating this equation over an arbitrary volume V of the body in $\mathscr{C}_t$ and using (1.9.8), this equation can be reduced to the form

$$\frac{d}{dt}\int_V \tfrac{1}{2}\rho v_i v_i \, dV + \int_V T_{ij}D_{ij} \, dV = \int_S t_i v_i \, dS + \int_V \rho b_i v_i \, dV \tag{1.13.5}$$

where again we have used the divergence theorem, (1.11.11), (1.11.18), and V is understood to be moving with a given set of particles of the body.

Equation (1.13.5) is known as the *mechanical energy-balance equation*. The terms on the left-hand side are the rate of increase of kinetic energy and the rate of working of the stresses, sometimes known as the *stress power*. The terms on the right-hand side represent the rate of working of the surface forces and body forces respectively.

In a full thermodynamic theory, the way in which the stress power contributes to an increase of stored energy and dissipation as heat has to be considered. In the present volume we neglect thermal effects entirely; the stress power is assumed to be absorbed completely in a stored energy depending only on the strain. This is explained further in Chapter 2.

1.14   Piola stresses

We see from (1.11.11) that the stress tensor **T** enables us to calculate stress vectors which are measured per unit area of surface in the current configuration. In some situations it is more convenient to be able to calculate the stress vector in $\mathscr{C}_t$ measured per unit area of the corresponding element of surface in $\mathscr{C}_0$. We first show how the area of an element of surface in $\mathscr{C}_t$ is related to its corresponding area in $\mathscr{C}_0$.

Consider two line elements $d\mathbf{X}$ and $d\mathbf{Y}$ at a point $\mathbf{X}$ in $\mathscr{C}_0$. These define a parallelogram-shaped element of surface with area dS_0 and a unit normal vector $\mathbf{N}$, where

$$d\mathbf{X} \times d\mathbf{Y} = \mathbf{N}\, dS_0 \qquad (1.14.1)$$

Likewise, the corresponding elements $d\mathbf{x}$ and $d\mathbf{y}$ in $\mathscr{C}_t$ define a surface element of area dS and with unit normal $\mathbf{n}$, where

$$d\mathbf{x} \times d\mathbf{y} = \mathbf{n}\, dS \qquad (1.14.2)$$

Using (1.3.2), we have therefore

$$n_i\, dS = \varepsilon_{ijk}\, dx_j\, dy_k = \varepsilon_{ijk} x_{j,L} x_{k,M}\, dX_L\, dY_M$$

Hence, from (1.14.1),

$$\begin{aligned} n_i x_{i,K}\, dS &= \varepsilon_{ijk} x_{i,K} x_{j,L} x_{k,M}\, dX_L\, dY_M \\ &= \varepsilon_{KLM} J\, dX_L\, dY_M, \text{ using (1.3.9)}, \\ &= J N_K\, dS_0 \end{aligned} \qquad (1.14.3)$$

Multiplying both sides by the inverse deformation gradients (defined by (1.4.18)), we may write (1.14.3) in the form

$$n_i\, dS = J X_{K,i} N_K\, dS_0 \qquad (1.14.4)$$

Now using (1.11.11) we see that the components of the force on the surface element dS in $\mathscr{C}_t$ can be written as

$$t_i\, dS = T_{ji} n_j\, dS = J X_{K,j} T_{ji} N_K\, dS_0 = S_{Ki} N_K\, dS_0 \qquad (1.14.5)$$

where

$$S_{Ki} = J X_{K,j} T_{ji}, \qquad \mathbf{S} = J \mathbf{F}^{-1} \mathbf{T} \qquad (1.14.6)$$

The S_{Ki} are the components of the first Piola stress tensor and, in general,

$$S_{Ki} \neq S_{iK}$$

However, a second Piola stress tensor may be defined which has components

$$\bar{S}_{KL} = X_{L,i} S_{Ki} \qquad (1.14.7)$$

and, because of the symmetry of T_{ij},

$$\bar{S}_{KL} = \bar{S}_{LK} \qquad (1.14.8)$$

so that the tensor $\bar{\mathbf{S}}$ is symmetric.

In view of (1.14.5) and (1.11.11), the principle of linear momentum (1.11.2) may be referred to the reference configuration and written in the form

$$\int_{S_0} S_{Ki} N_K \, dS_0 + \int_{V_0} \rho_0 b_i \, dV_0 = \frac{d}{dt} \int_{V_0} \rho_0 v_i \, dV_0.$$

The point-form then becomes

$$S_{Ki,K} + \rho_0 b_i = \rho_0 \frac{Dv_i}{Dt} \tag{1.14.9}$$

From the principle of angular momentum (1.11.15), we deduced the symmetry of T_{ij}, and so from (1.14.6) we see that

$$S_{Ki} x_{j,K} = S_{Kj} x_{i,K}, \qquad \mathbf{S}^T \mathbf{F}^T = \mathbf{F} \mathbf{S} \tag{1.14.10}$$

It is perhaps useful at this point to illustrate the physical significance of the components S_{Ki} of the first Piola stress tensor. Suppose that the element of surface dS_0 in $\mathscr{C}_0$ has a normal in the X_1-direction, so that $\mathbf{N} = (1, 0, 0)$ (see Fig. 1.11). Then, using (1.14.5),

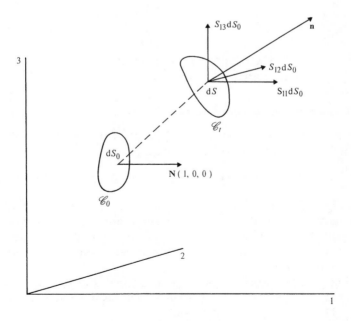

Fig. 1.11

we see that the force on the corresponding element in $\mathscr{C}_t$ has components $S_{1i}\,dS_0$. Similar reasoning gives an interpretation of S_{2i} and S_{3i}. Hence, S_{Ki} represents the component in the i-direction of the force acting on an element of surface in $\mathscr{C}_t$ which corresponds to an element in $\mathscr{C}_0$ with normal in the K-direction, the force being expressed per unit area of the element in $\mathscr{C}_0$.

Example 1.14.1

Express the Cauchy stress components in terms of the first Piola stress components when the deformation is given by (1.5.5) and (1.5.12), respectively.

Explain the physical significance of S_{11} and S_{12} when the deformation is given by (1.5.12).

Example 1.14.2

Let $\mathbf{s(N)}$ denote the force on an element of surface in $\mathscr{C}_t$ measured per unit area of the corresponding element with unit normal $\mathbf{N}$ in $\mathscr{C}_0$. Show that

$$s_i = S_{Ki}N_K \tag{1.14.11}$$

1.15 Cylindrical and spherical polar coordinates

(i) Cylindrical polar coordinates

The cylindrical polar coordinates (r, θ, z) of the point $\mathbf{x}$ in $\mathscr{C}_t$ are related to its Cartesian coordinates (x_1, x_2, x_3) by

$$x_1 = r\cos\theta, \quad x_2 = r\sin\theta, \quad x_3 = z, \qquad r \geqslant 0, \quad 0 \leqslant \theta < 2\pi \tag{1.15.1}$$

(see Fig. 1.12). The coordinate surfaces are the coaxial cylinders $r = $ constant, the half-planes $\theta = $ constant, and the planes $z = $ constant. For points $\mathbf{X}$ in the reference configuration, we use cylindrical polar coordinates (R, Θ, Z) related similarly to the Cartesian coordinates (X_1, X_2, X_3).

Unit vectors associated with the coordinates (r, θ, z) in $\mathscr{C}_t$ are denoted by $\mathbf{e}_r$, $\mathbf{e}_\theta$, and $\mathbf{e}_z$, and their directions are indicated in Fig. 1.12. The stress vectors acting on the coordinate surfaces may be

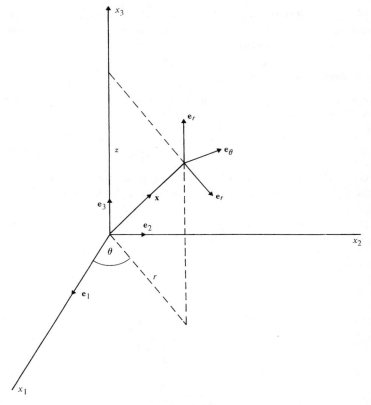

Fig. 1.12

resolved in the directions of these unit vectors as follows:

$$\mathbf{t}(\mathbf{e}_r) = T_{rr}\mathbf{e}_r + T_{r\theta}\mathbf{e}_\theta + T_{rz}\mathbf{e}_z$$
$$\mathbf{t}(\mathbf{e}_\theta) = T_{\theta r}\mathbf{e}_r + T_{\theta\theta}\mathbf{e}_\theta + T_{\theta z}\mathbf{e}_z \qquad (1.15.2)$$
$$\mathbf{t}(\mathbf{e}_z) = T_{zr}\mathbf{e}_r + T_{z\theta}\mathbf{e}_\theta + T_{zz}\mathbf{e}_z$$

The coefficients T_{rr}, $T_{r\theta}$, etc., again form a symmetric array and are known as the components of the Cauchy stress tensor in cylindrical polar coordinates.

Example 1.15.1

Write down the direction cosines representing the rotation from the triad $\mathbf{e}_1$, $\mathbf{e}_2$, $\mathbf{e}_3$ to $\mathbf{e}_r$, $\mathbf{e}_\theta$, $\mathbf{e}_z$.

Find the components of the Cauchy stress tensor in cylindrical coordinates at the point (r, θ, z) corresponding to a uniaxial tension T in the x_1-direction.

(ii) Spherical polar coordinates

The spherical polar coordinates (r, θ, ϕ) of the point $\mathbf{x}$ in $\mathscr{C}_t$ are related to its Cartesian coordinates (x_1, x_2, x_3) by

$$x_1 = r \sin \theta \cos \phi, \quad x_2 = r \sin \theta \sin \phi, \quad x_3 = r \cos \theta$$
$$r \geq 0, \quad 0 \leq \theta \leq \pi, \quad 0 \leq \phi < 2\pi \tag{1.15.3}$$

(see Fig. 1.13). The symbols r, θ employed here are different from those appearing in (1.15.1); however, the context will always

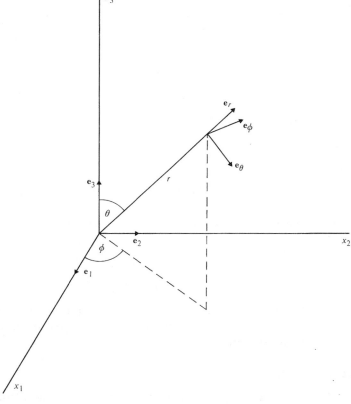

Fig. 1.13

make clear which coordinate system is being used. Spherical polar coordinates (R, Θ, Φ) are related similarly to the Cartesian coordinates (X_1, X_2, X_3) of the point $\mathbf{X}$ in the reference configuration.

The unit vectors $\mathbf{e}_r$, $\mathbf{e}_\theta$, and $\mathbf{e}_\phi$ associated with the coordinates (r, θ, ϕ) are shown in Fig. 1.13. The coordinate surfaces are the spheres $r = $ constant, the cones $\theta = $ constant, and the half-planes $\phi = $ constant. The stress vectors acting on these coordinate surfaces may be expressed as follows:

$$\mathbf{t}(\mathbf{e}_r) = T_{rr}\mathbf{e}_r + T_{r\theta}\mathbf{e}_\theta + T_{r\phi}\mathbf{e}_\phi$$
$$\mathbf{t}(\mathbf{e}_\theta) = T_{\theta r}\mathbf{e}_r + T_{\theta\theta}\mathbf{e}_\theta + T_{\theta\phi}\mathbf{e}_\phi \qquad (1.15.4)$$
$$\mathbf{t}(\mathbf{e}_\phi) = T_{\phi r}\mathbf{e}_r + T_{\phi\theta}\mathbf{e}_\theta + T_{\phi\phi}\mathbf{e}_\phi$$

The symmetric array of coefficients T_{rr}, $T_{r\theta}$, etc., are the components of the Cauchy stress tensor in spherical polar coordinates.

Examples 1

1. The motion of a continuous body is given by

$$x_1 = X_1 \cos \Omega t - X_2 \sin \Omega t$$
$$x_2 = X_1 \sin \Omega t + X_2 \cos \Omega t, \quad x_3 = X_3$$

where Ω is a positive constant. Describe this motion geometrically and find the material and spatial descriptions of the velocity and acceleration fields.

2. Find the value of J and the material and spatial descriptions of the velocity field for the motion given by

$$x_1 = \alpha(t)X_1, \quad x_2 = \beta(t)X_2, \quad x_3 = \gamma(t)X_3$$

where α, β, and γ are positive-valued functions of time. Calculate $\partial \mathbf{v}/\partial t$, $D\mathbf{v}/Dt$, and $(\mathbf{v} \cdot \nabla)\mathbf{v}$ and hence verify the relation (1.2.4). When $\alpha(t) = e^{\lambda t}$, $\beta(t) = \gamma(t) = 1$ ($\lambda = $ constant), show that $\partial \mathbf{v}/\partial t = 0$ but $D\mathbf{v}/Dt \neq 0$.

3. Take the material time derivative of (1.4.17) to show that

$$\frac{D}{Dt}((dl)^2) = D_{ij}\, dx_i\, dx_j$$

4. In the case of the simple shear deformation (1.5.12), show that the principal stretches are $\{1+\frac{1}{2}\kappa^2 \pm \kappa(1+\frac{1}{4}\kappa^2)^{\frac{1}{2}}\}^{\frac{1}{2}}$ and 1. Find the principal axes of the strain tensors **B** and **C**.

5. A body undergoes the deformation

$$x_1 = f(X_1)\cos(\alpha X_2), \quad x_2 = f(X_1)\sin(\alpha X_2), \quad x_3 = X_3$$

where $f(X_1)$ is a positive, differentiable increasing function of X_1 and α is a constant. Show that under this deformation, planes which are initially perpendicular to the X_1-axis are deformed into cylindrical surfaces. Find the surfaces into which the planes which are initially perpendicular to the X_2-axis are deformed. If the body initially occupies the region

$$a \leqslant X_1 \leqslant b, \quad -c \leqslant X_2 \leqslant c, \quad 0 \leqslant X_3 \leqslant d \quad (a>0)$$

sketch the shape of that part of the current configuration which lies in the plane $x_3 = 0$.

Finally, determine the form of the function $f(X_1)$ for the deformation to be isochoric.

6. A body undergoes the deformation

$$x_1 = (1+\alpha)X_1 + \beta X_2, \quad x_2 = \beta X_1 + (1+\alpha)X_2, \quad x_3 = X_3$$

where α and β are positive constants such that $1 + \alpha > \beta$. If ρ_0 denotes the density in the reference configuration, show that after this deformation

$$\rho = \rho_0 \{(1+\alpha)^2 - \beta^2\}^{-1}$$

and that the planes which in the undeformed configuration are given by

$$X_1 = a, \qquad X_2 = b$$

where a and b are positive constants, are deformed into the planes

$$\beta x_2 = (1+\alpha)x_1 - a\{(1+\alpha)^2 - \beta^2\}$$
$$(1+\alpha)x_2 = \beta x_1 + b\{(1+\alpha)^2 - \beta^2\}$$

respectively.

In the undeformed configuration, a line element PQ of length dL lies in the plane $X_3 = 0$ and is inclined at an angle θ to the X_1 axis. Find the length dl of this element after the deformation. Show that, for the special case when $\alpha = \beta$, the planes

$$X_1 + X_2 = \text{constant}$$

in the undeformed configuration are unstretched by the deformation.

7. Give a geometrical description of the deformation

$$x_1 = X_1 - \tau X_2 X_3, \quad x_2 = X_2 + \tau X_1 X_3, \quad x_3 = X_3$$

where τ is a constant, and calculate the components of **F** and **C**. Is the deformation isochoric?

Find the surface into which the cylinder $X_1^2 + X_2^2 = a^2$ deforms.

8. If axes $Ox_1'x_2'x_3'$ are obtained from $Ox_1x_2x_3$ by a rotation through an angle α about Ox_3 using the usual corkscrew rule for the sign of the angle, and if T_{ij}' and T_{ij} are the components of the stress tensor with respect to these sets of axes, show that

$$T_{11}' + T_{22}' = T_{11} + T_{22}$$
$$T_{11}' - T_{22}' + 2iT_{12}' = e^{-2i\alpha}(T_{11} - T_{22} + 2iT_{12})$$
$$T_{13}' + iT_{23}' = e^{-i\alpha}(T_{13} + iT_{23})$$

9. The stress tensor at a certain point has components

$$\begin{pmatrix} 3 & 1 & 1 \\ 1 & 0 & 2 \\ 1 & 2 & 0 \end{pmatrix}$$

in a given system of units. Find the stress vector on an element of surface with normal in the direction of the vector $(0, 1, 1)$. Give the normal and shear components of this stress vector.

Find the principal stresses and the direction cosines of the principal axes.

10. The stress tensor at a certain point has components

$$\begin{pmatrix} 0 & 0 & -2\beta \\ 0 & 0 & \beta \\ -2\beta & \beta & 0 \end{pmatrix}$$

where β is a constant. Show that the principal stresses are 0 and $\pm\beta\sqrt{5}$. Find the direction for which the principal stress is $\beta\sqrt{5}$.

11. The distribution of stress in a square slab which occupies the region

$$|x_1| \leqslant a, \quad |x_2| \leqslant a, \quad |x_3| \leqslant h$$

is given by

$$T_{11} = -p(x_1^2 - x_2^2)/a^2, \quad T_{22} = p(x_1^2 - x_2^2)/a^2, \quad T_{33} = 0$$
$$T_{12} = 2px_1x_2/a^2, \quad T_{23} = T_{31} = 0.$$

No body forces act on the slab. Verify that the slab is in equilibrium and calculate the components of the surface traction on each of its faces. Show that the resultant force which acts on each of the faces $x_1 = \pm a$, $x_2 = \pm a$ has magnitude $8pah/3$, and indicate on a sketch the lines of action of these resultants.

[*Hint:* You may find the comments on boundary conditions in Section 3.1 useful in this problem.]

12. The stress tensor in a material which is at rest in a certain frame of reference has components

$$\tfrac{1}{4}\rho\omega^2 \begin{pmatrix} x_1^2 & 2x_1x_2 & 0 \\ 2x_1x_2 & x_2^2 & 0 \\ 0 & 0 & 2(x_1^2 + x_2^2) \end{pmatrix}$$

where ω is a constant. Find the body force which must be acting in that frame.

13. A body not subject to body forces is in equilibrium under a state of stress given by

$$T_{ij} = Ta_ia_j$$

where the scalar T may vary with position but $\mathbf{a}$ is a *constant* unit vector. Show that grad T must be everywhere perpendicular to $\mathbf{a}$. Prove that the stress vector acting across an area with arbitrary unit normal $\mathbf{n}$ is parallel to $\mathbf{a}$. Evaluate this vector when: (i) $\mathbf{n}$ is parallel to $\mathbf{a}$; (ii) $\mathbf{n}$ is perpendicular to $\mathbf{a}$.

14. A body in equilibrium in the absence of body forces occupies a region R which does not include the origin of coordinates O. If

the stress components at a typical point P in R, with coordinates x_i, are of the form

$$T_{ij} = Tn_i n_j$$

where $T = T(x_1, x_2, x_3)$ is a function of position and $n(x_1, x_2, x_3)$ is a unit vector in the direction $\overrightarrow{OP}$, show that

$$x_j \frac{\partial T}{\partial x_j} + 2T = 0.$$

Evaluate T if it is a function only of distance from O.

15. Show that

$$\bar{S}_{KL} \frac{DC_{KL}}{Dt} = 2S_{Ki} v_{i,K}$$

Using this result, equation (1.14.9) and the divergence theorem show that referred to the reference configuration, the mechanical energy-balance equation takes the form

$$\frac{d}{dt} \int_{V_0} \tfrac{1}{2}\rho_0 v_i v_i \, dV_0 + \frac{1}{2} \int_{V_0} \bar{S}_{KL} \frac{DC_{KL}}{Dt} \, dV_0$$

$$= \int_{V_0} \rho_0 b_i v_i \, dV_0 + \int_{S_0} v_i S_{Ki} N_K \, dS_0$$

Finite elasticity: constitutive theory

2.1 Constitutive equations

The continuity equation, the principles of linear and angular momentum, and the energy equation, discussed in Chapter 1, hold for most materials regardless of their constitution. However, unless the body can be regarded as rigid, these equations are in general insufficient to determine the motion produced by given boundary conditions and body force. They need to be supplemented by a further set of equations, known as *constitutive equations*, which characterise the constitution of the body. Such a set of constitutive equations usually serves to define an ideal material, and much of the work on modern continuum mechanics has been concerned with the formulation of constitutive equations to model as closely as possible the behaviour of real materials.

For some ideal materials, the most convenient form for the constitutive equations is that in which the stress components at a particle are determined explicitly by the motion in the neighbourhood of that particle. However, a certain characteristic of the constitution of a material may sometimes be best described by a geometrical restriction on the class of motions of which the material is capable. Such a restriction is usually referred to as a *constraint*. The most common constraint, and indeed the only one we shall consider here, is that of incompressibility. Some materials can sustain large deformations without appreciable volume change, and we idealize this property by postulating the constraint

$$J = 1, \qquad (2.1.1)$$

or, equivalently,

$$v_{i,i} = D_{ii} = \operatorname{tr} \mathbf{D} = 0 \qquad (2.1.2)$$

In other words, only isochoric motions are possible in such materials. When no such explicit geometrical restriction is placed on the possible motions, the material is called *unconstrained*. As we consider only the constraint of incompressibility, from now on we refer to unconstrained materials as *compressible*.

In this chapter we discuss separately constitutive equations which are appropriate to two classes of ideal materials known as *compressible elastic materials* and *incompressible elastic materials*.

A convenient starting point for the theory of compressible elastic materials is to postulate a set of constitutive equations in which the stress components are regarded as single-valued functions of the deformation-gradient tensor, that is,

$$T_{ij} = f_{ij}(F_{kA}), \qquad \mathbf{T} = \mathbf{f}(\mathbf{F}) \tag{2.1.3}$$

where, to satisfy (1.11.18), the functions f_{ij} are chosen so that $f_{ij} = f_{ji}$.

We see from (1.3.2) that $\mathbf{F}$ gives a complete description of the deformation of an infinitesimal element of material at the point $\mathbf{X}$, and so, physically, (2.1.3) expresses the fact that the stress components at a particle $\mathbf{X}$ at time t are determined by the change of shape and orientation of an arbitrarily small element about the point $\mathbf{X}$.

The theory of elasticity based on (2.1.3) is usually referred to as *Cauchy elasticity*. Now for most of our development in this text we make an additional assumption: we suppose that the whole of the stress power (see equation (1.13.5)) is absorbed into, or derived from, a strain-energy function W depending only on $\mathbf{F}$; that is, we suppose that there exists a single-valued function $W(\mathbf{F})$ such that

$$\frac{d}{dt} \int_V \frac{\rho}{\rho_0} W \, dV = \int_V T_{ij} D_{ij} \, dV \tag{2.1.4}$$

The factor ρ/ρ_0 is introduced into the first integrand as it is more convenient to define W to be the strain-energy function for the body in $\mathscr{C}_t$, measured per unit volume of the corresponding region of $\mathscr{C}$. Using (1.9.8) and the fact that equation (2.1.4) holds for all regions V within the body, provided the integrand is continuous, the point form of (2.1.4) is

$$\frac{\rho}{\rho_0} \frac{DW}{Dt} = T_{ij} D_{ij} \tag{2.1.5}$$

The theory of elasticity incorporating this further assumption is known as *Green elasticity* or *hyperelasticity*. Some of the results obtained below hold for both Cauchy and Green elasticity, but the latter is more easily related to the experimental results.

We note here that the use of $\mathbf{F}$ as the only independent variable in the constitutive equations means that W and $\mathbf{T}$ are completely determined when $\mathscr{C}$ and $\mathscr{C}_t$ are specified. The rate at which the deformation is taking place or the intermediate configurations have no effect on the stress. To develop theories in which rate effects, such as viscosity, are present, some dependence on the rate-of-deformation tensor $\mathbf{D}$ would have to be taken into account.

Example 2.1.1

Let $\mathbf{F}_0$ be an arbitrary constant tensor for which $\det \mathbf{F}_0 > 0$. Show that in a compressible elastic body there exists a motion for which the deformation-gradient tensor $\mathbf{F} = \mathbf{F}_0$ throughout the body at time $t = t_0$, and for which the velocity-gradient tensor $\mathbf{L}$ with components $(v_{i,j})$ takes an arbitrary constant value. Find the body-force field required to produce such a motion.

Consider the motion

$$x_i = A_{iK}(t)X_K \tag{2.1.6}$$

where $\det \mathbf{A} > 0$. In this case, the reference configuration $\mathscr{C}$ is not necessarily a configuration through which the body passes during the motion. The body passes through $\mathscr{C}$ if and only if $\mathbf{A}(t) = \mathbf{I}$ at some time t.

The deformation-gradient tensor is

$$\mathbf{F} = \mathbf{A}(t) \tag{2.1.7}$$

and we choose

$$\mathbf{A}(t_0) = \mathbf{F}_0 \tag{2.1.8}$$

Now the velocity gradients may be written

$$v_{i,j} = v_{i,K}X_{K,j} = \frac{\mathrm{D}}{\mathrm{D}t}(x_{i,K})X_{K,j}$$

or, if $\mathbf{L}$ denotes the tensor with components $(v_{i,j})$,

$$\mathbf{L} = \dot{\mathbf{F}}\mathbf{F}^{-1} \tag{2.1.9}$$

where the superposed dot denotes the material time derivative.

Hence, using (2.1.7),

$$\dot{\mathbf{A}} = \mathbf{L}\mathbf{A} \tag{2.1.10}$$

We now suppose that $\mathbf{L}$ is an arbitrary given constant tensor and we choose $\mathbf{A}$ to be the solution of the differential equation (2.1.10) subject to the condition (2.1.8). (It is easy to verify that the solution is

$$\mathbf{A} = \mathbf{F}_0\left(\mathbf{I} + \frac{\mathbf{L}(t-t_0)}{1!} + \frac{\mathbf{L}^2(t-t_0)^2}{2!} + \ldots\right)$$

sometimes denoted by

$$\mathbf{A} = \mathbf{F}_0 e^{\mathbf{L}(t-t_0)} \tag{2.1.11}$$

The motion (2.1.6) with this value of $\mathbf{A}$ has the required properties.

The deformation-gradient tensor $\mathbf{F}$ depends only on t, so that in a compressible elastic material $\mathbf{T}$ depends only on t. Hence, using (1.11.14), the body force required to produce this motion is given by

$$b_i = \frac{Dv_i}{Dt} = \ddot{A}_{iK}X_K = L_{ij}^2 A_{jK}X_K \tag{2.1.12}$$

2.2 Invariance under superposed rigid-body motions

In Chapter 1 we discussed how the motion of the body could be described by giving the position $\mathbf{x}$ of each particle $\mathbf{X}$ at time t in the form of an equation

$$\mathbf{x} = \chi(\mathbf{X}, t) \tag{2.2.1}$$

Consider now another motion of the same body in which each particle is in the same position as in the motion (2.2.1) *relative to the rest of the body*, but in which the body, as a whole, has been rotated and translated, at each instant of time, as a rigid body from its position in $\mathscr{C}_t$. In other words, the current configuration $\hat{\mathscr{C}}_t$ of the body in this new motion is obtained by a time-dependent, rigid-body rotation and translation from $\mathscr{C}_t$. Thus the position $\hat{\mathbf{x}}$ of the particle $\mathbf{X}$ in $\hat{\mathscr{C}}_t$ is given by

$$\hat{x}_i = Q_{ij}(t)\chi_j(\mathbf{X}, t) + c_i(t) \tag{2.2.2}$$

where $\mathbf{Q}(t)$ is a proper orthogonal, time-dependent tensor so that

$$\mathbf{Q}\mathbf{Q}^{\mathrm{T}} = \mathbf{Q}^{\mathrm{T}}\mathbf{Q} = \mathbf{I}, \qquad \det \mathbf{Q} = +1 \qquad (2.2.3)$$

and $\mathbf{c}(t)$ is a time-dependent vector.

Now in each motion the particles occupy the same relative positions. We assume that W is affected only by these relative positions not by their absolute positions in space. Thus the value of W at a particle $\mathbf{X}$ is the same in both motions at each instant of time. Underlying this assumption is an idea known as the *principle of material frame-indifference* since, mathematically, our statement is identical to the requirement that W should be invariant under transformations of (2.2.1) to arbitrary moving right-handed reference frames. Physically, the latter means that W is independent of the motion of the observer. We are, of course, omitting all relativistic effects from our considerations.

The components of the deformation gradient associated with (2.2.2) are

$$\hat{F}_{iA} = \hat{x}_{i,A} \equiv \partial \hat{x}_i / \partial X_A \qquad (2.2.4)$$

Now from (2.2.2)

$$\hat{x}_{i,A} = Q_{ij} x_{j,A} \qquad (2.2.5)$$

giving

$$\hat{\mathbf{F}} = \mathbf{Q}\mathbf{F} \qquad (2.2.6)$$

2.3 Invariance of the strain energy under superposed rigid-body motions

The value of the strain energy associated with the motion (2.2.1) is $W(\mathbf{F})$ whereas that associated with (2.2.2) is $W(\hat{\mathbf{F}})$. Thus our earlier statement that the strain-energy function is unaffected by a rigid rotation of the body now takes the form

$$W(\mathbf{F}) = W(\hat{\mathbf{F}}) = W(\mathbf{Q}\mathbf{F}) \qquad (2.3.1)$$

where (2.3.1) must hold for *all* proper orthogonal $\mathbf{Q}$. Thus we see that W cannot depend upon $\mathbf{F}$ in an arbitrary manner but only in such a way that (2.3.1) is satisfied. Using the polar decomposition theorem (equation $(1.4.2)_1$), (2.3.1) becomes

$$W(\mathbf{F}) = W(\mathbf{Q}\mathbf{R}\mathbf{U}) \qquad (2.3.2)$$

Since this must hold for *all* proper orthogonal $\mathbf{Q}$ it must hold when $\mathbf{Q}$ takes the value of $\mathbf{R}^{\mathrm{T}}$ at any given point $\mathbf{X}$ in which case we have

$$W(\mathbf{F}) = W(\mathbf{R}^{\mathrm{T}}\mathbf{R}\mathbf{U}) = W(\mathbf{U}) \tag{2.3.3}$$

in view of $(1.4.3)_1$.

Now $\mathbf{U}$ is a positive-definite tensor, and is therefore uniquely determined by its square, $\mathbf{U}^2 = \mathbf{C}$. This is perhaps most easily understood by choosing axes so that $\mathbf{U}$ and $\mathbf{C}$ are diagonal. Then, if (u_1, u_2, u_3) and (c_1, c_2, c_3) are the respective diagonal elements of $\mathbf{U}$ and $\mathbf{C}$, $c_1 = u_1^2$, $c_2 = u_2^2$, $c_3 = u_3^2$ and the set (c_1, c_2, c_3) uniquely determines (u_1, u_2, u_3) since each of the latter set must be positive. This means that, instead of writing W as a function of $\mathbf{U}$, it may be regarded as a function of $\mathbf{C}$:

$$W(\mathbf{F}) = \tilde{W}(\mathbf{C}) \tag{2.3.4}$$

where for clarity we are now introducing new functional symbols as required.

We can also show that (2.3.4) is a sufficient condition for (2.3.1) to hold. Now

$$
\begin{aligned}
W(\hat{\mathbf{F}}) &= \tilde{W}(\hat{\mathbf{C}}) \\
&= \tilde{W}(\hat{\mathbf{F}}^{\mathrm{T}}\hat{\mathbf{F}}) \qquad \text{by} \quad (1.4.8)_1 \\
&= \tilde{W}((\mathbf{Q}\mathbf{F})^{\mathrm{T}}(\mathbf{Q}\mathbf{F})) \qquad \text{using} \quad (2.2.6) \\
&= \tilde{W}(\mathbf{F}^{\mathrm{T}}\mathbf{Q}^{\mathrm{T}}\mathbf{Q}\mathbf{F}) \\
&= \tilde{W}(\mathbf{F}^{\mathrm{T}}\mathbf{F}) \qquad \text{using} \quad (2.2.3)_1 \\
&= \tilde{W}(\mathbf{C}) \\
&= W(\mathbf{F}) \qquad \text{using} \quad (2.3.4)
\end{aligned}
$$

Thus a necessary and sufficient condition for the relation (2.3.1) to be satisfied is that W depends on $\mathbf{F}$ only through $\mathbf{C}$.

2.4   The stress tensor in terms of the strain-energy function

Using (2.3.4) and the chain rule for partial differentiation we have

$$\frac{\mathrm{D}W}{\mathrm{D}t} = \frac{\partial W}{\partial C_{\mathrm{PQ}}} \frac{\mathrm{D}}{\mathrm{D}t}(C_{\mathrm{PQ}}) \tag{2.4.1}$$

and from the definition of **C** (equation $(1.4.8)_1$)

$$\frac{D}{Dt}(C_{PQ}) = \frac{D}{Dt}(x_{i,P}x_{i,Q})$$

$$= x_{i,P}\frac{D}{Dt}(x_{i,Q}) + x_{i,Q}\frac{D}{Dt}(x_{i,P})$$

$$= x_{i,P}v_{i,Q} + x_{i,Q}v_{i,P} \quad \text{using} \quad (1.2.2)$$

$$= x_{i,P}v_{i,j}x_{j,Q} + v_{i,j}x_{j,P}x_{i,Q}$$

where now the velocity field is regarded as a function of **x** and t. Thus

$$\frac{D}{Dt}(C_{PQ}) = (x_{i,P}x_{j,Q} + x_{j,P}x_{i,Q})v_{i,j} \tag{2.4.2}$$

Since $T_{ij}D_{ij} = T_{ij}v_{i,j}$, combining (2.4.1) and (2.4.2), equation (2.1.5) may be rewritten as

$$P_{ij}v_{i,j} = 0 \tag{2.4.3}$$

where

$$P_{ij} = T_{ij} - \frac{\rho}{\rho_0}(x_{i,P}x_{j,Q} + x_{j,P}x_{i,Q})\frac{\partial W}{\partial C_{PQ}} \tag{2.4.4}$$

Now P_{ij} is completely determined by the values of $x_{i,A}$ and it can be shown that (see Example 2.1.1), for any given set of values of $x_{i,A}$ at any given point and any instant of time, there exists a class of possible motions of the body for which the velocity gradients $v_{i,j}$ are arbitrary. Since (2.4.3) must hold for every such motion it follows that

$$P_{ij} = 0 \tag{2.4.5}$$

Hence, using (2.4.4),

$$T_{ij} = \frac{\rho}{\rho_0}(x_{i,P}x_{j,Q} + x_{j,P}x_{i,Q})\frac{\partial W}{\partial C_{PQ}}$$

$$= \frac{\rho}{\rho_0}x_{i,P}x_{j,Q}\left(\frac{\partial W}{\partial C_{PQ}} + \frac{\partial W}{\partial C_{QP}}\right) \tag{2.4.6}$$

Since **C** is symmetric, W may be expressed symmetrically in the suffixes of C_{AB} and (2.4.6) may be written

$$T_{ij} = 2\frac{\rho}{\rho_0}x_{i,P}x_{j,Q}\frac{\partial W}{\partial C_{PQ}} \tag{2.4.7}$$

The reader should note, for example, that if W may be expressed as

$$W = C_{11}^2 + C_{12}^2$$

then

$$\frac{\partial W}{\partial C_{12}} = 2C_{12} \quad \text{and} \quad \frac{\partial W}{\partial C_{21}} = 0$$

and so

$$\frac{\partial W}{\partial C_{12}} \neq \frac{\partial W}{\partial C_{12}}$$

However, when W is expressed symmetrically,

$$W = C_{11}^2 + \tfrac{1}{4}(C_{12} + C_{21})^2$$

and

$$\frac{\partial W}{\partial C_{12}} = \frac{\partial W}{\partial C_{21}}$$

Using (1.9.4), (1.14.6), and (2.4.6), the components of the first Piola stress tensor are given by

$$S_{Ki} = x_{i,L}\left(\frac{\partial W}{\partial C_{KL}} + \frac{\partial W}{\partial C_{LK}}\right) \qquad (2.4.8)$$

Equations (2.4.6), (2.4.7), or (2.4.8) are the constitutive equations for the stress in a compressible elastic material.

2.5 Material symmetry. Strain-energy function for an isotropic material

So far, no assumptions have been made about the symmetry which the material may possess. We now consider an important kind of symmetry which reflects the physical idea that, when the material is in its reference configuration, no preferred directions can be detected by the response of the material in any stress–strain experiment. Materials such as wood, for example, would not have this property since the direction of the grain could easily be detected by experiment. But rubber can be processed so that its response to a given state of stress is largely independent of the orientation of its reference configuration.

Consider a reference configuration $\mathscr{C}$ in which the particles are labelled by their position vectors relative to a fixed set of axes. Then under the motion (1.1.1) the particle $\mathbf{X}$ is moved to the position $\mathbf{x}$ where

$$\mathbf{x} = \chi(\mathbf{X}, t) \tag{2.5.1}$$

and the strain energy associated with this motion is given by (2.3.4). Now consider a second configuration $\bar{\mathscr{C}}$ in which the point in space which in $\mathscr{C}$ was occupied by the particle $\mathbf{X}$ is now occupied by the particle labelled $\bar{\mathbf{X}}$, where

$$\bar{\mathbf{X}} = \mathbf{Q}\mathbf{X} \tag{2.5.2}$$

and $\mathbf{Q}$ is a constant *orthogonal* tensor. Since the particle $\bar{\mathbf{X}}$ occupies the same position in $\bar{\mathscr{C}}$ as the particle $\mathbf{X}$ did in $\mathscr{C}$ under the motion χ, $\bar{\mathbf{X}}$ will occupy the position $\mathbf{x}$, that is

$$\mathbf{x} = \chi(\bar{\mathbf{X}}, t) \tag{2.5.3}$$

The strain energy associated with (2.5.3) is given by an expression of the form (2.3.4) but the deformation gradients are now calculated relative to $\bar{\mathscr{C}}$ so that

$$W = \tilde{W}(\bar{\mathbf{C}}) \tag{2.5.4}$$

where

$$\bar{\mathbf{C}} = \bar{\mathbf{F}}^{\mathrm{T}}\bar{\mathbf{F}} \tag{2.5.5}$$

and in components

$$\bar{F}_{iA} = \partial x_i / \partial \bar{X}_A \tag{2.5.6}$$

In general, the value of the strain energy given by (2.3.4) and (2.5.4) need not be the same. However, if the values are the same for all $\mathbf{Q}$ then the body is said to be *isotropic* relative to the reference configuration $\mathscr{C}$, in which case

$$\tilde{W}(\mathbf{C}) = \tilde{W}(\bar{\mathbf{C}}) \tag{2.5.7}$$

A material which is not isotropic is known as *anisotropic*.

Now

$$F_{iB} = \frac{\partial x_i}{\partial X_B} = \frac{\partial x_i}{\partial \bar{X}_A} \frac{\partial \bar{X}_A}{\partial X_B} = \frac{\partial x_i}{\partial \bar{X}_A} Q_{AB} = \bar{F}_{iA} Q_{AB} \tag{2.5.8a}$$

or

$$\mathbf{F} = \bar{\mathbf{F}}\mathbf{Q} \tag{2.5.8b}$$

so that

$$\bar{\mathbf{C}} = \bar{\mathbf{F}}^{\mathrm{T}}\bar{\mathbf{F}} = (\mathbf{F}\mathbf{Q}^{\mathrm{T}})^{\mathrm{T}}(\mathbf{F}\mathbf{Q}^{\mathrm{T}}) = \mathbf{Q}\mathbf{F}^{\mathrm{T}}\mathbf{F}\mathbf{Q}^{\mathrm{T}} = \mathbf{Q}\,\mathbf{C}\mathbf{Q}^{\mathrm{T}} \qquad (2.5.9)$$

Thus for an isotropic material we require

$$\tilde{W}(\mathbf{C}) = \tilde{W}(\mathbf{Q}\mathbf{C}\mathbf{Q}^{\mathrm{T}}) \qquad (2.5.10)$$

for all orthogonal tensors $\mathbf{Q}$. If (2.5.10) is true then it follows that (see Spencer (1980) Section 10.2).

$$W = \breve{W}(I_1, I_2, I_3) \qquad (2.5.11)$$

where I_1, I_2, I_3 are the principal invariants of $\mathbf{C}$ or $\mathbf{B}$ defined by (1.4.23).

2.6 The stress tensor for an isotropic material

Since (2.5.11) is a symmetric function of $\mathbf{C}$, using (2.4.7) the stress tensor is given in component form by

$$T_{ij} = \frac{2}{I_3^{\frac{1}{2}}} x_{i,P} x_{j,Q} \left\{ W_1 \frac{\partial I_1}{\partial C_{PQ}} + W_2 \frac{\partial I_2}{\partial C_{PQ}} + W_3 \frac{\partial I_3}{\partial C_{PQ}} \right\} \qquad (2.6.1)$$

where

$$W_1 \equiv \partial W/\partial I_1, \quad W_2 \equiv \partial W/\partial I_2, \quad W_3 \equiv \partial W/\partial I_3 \qquad (2.6.2)$$

and we have used (1.9.4) as well as the result $J = \det \mathbf{F} = (\det \mathbf{C})^{\frac{1}{2}} = I_3^{\frac{1}{2}}$. Now

$$\frac{\partial I_1}{\partial C_{PQ}} = \frac{\partial}{\partial C_{PQ}}(C_{KK}) = \delta_{PQ}$$

$$\frac{\partial I_2}{\partial C_{PQ}} = \frac{1}{2} \frac{\partial}{\partial C_{PQ}}(C_{KK}C_{LL} - C_{KL}C_{KL}) = I_1 \delta_{PQ} - C_{PQ} \qquad (2.6.3)$$

Applying the Cayley–Hamilton theorem to $\mathbf{C}$ (cf. (1.4.29)), we obtain

$$C_{PR}C_{RS}C_{SQ} - I_1 C_{PR}C_{RQ} + I_2 C_{PQ} - I_3 \delta_{PQ} = 0 \qquad (2.6.4)$$

so that contracting over P and Q this gives

$$3I_3 = C_{PR}C_{RS}C_{SP} - I_1 C_{PR}C_{RP} + I_2 C_{PP} \qquad (2.6.5)$$

Differentiating both sides of (2.6.5) with respect to C_{PQ} and using

(2.6.3), it follows that

$$\frac{\partial I_3}{\partial C_{PQ}} = I_2 \delta_{PQ} - I_1 C_{PQ} + C_{PR} C_{RQ} \qquad (2.6.6)$$

Also recalling the definition of the left Cauchy–Green tensor **B** (equation $(1.4.8)_2$)

$$x_{i,P} x_{j,Q} \delta_{PQ} = x_{i,P} x_{j,P} = B_{ij}$$

$$x_{i,P} x_{j,Q} C_{PQ} = x_{i,P} x_{j,Q} x_{l,P} x_{l,Q} = B_{il} B_{lj}$$

$$x_{i,P} x_{j,Q} C_{PR} C_{RQ} = x_{i,P} x_{j,Q} x_{l,P} x_{l,R} x_{k,R} x_{k,Q}$$

$$= B_{il} B_{lk} B_{kj}$$

$$(2.6.7)$$

Thus

$$x_{i,P} x_{j,Q} \frac{\partial I_3}{\partial C_{PQ}} = I_2 B_{ij} - I_1 B_{il} B_{lj} + B_{il} B_{lk} B_{kj}$$

$$= I_2 B_{ij} - I_1 B_{il} B_{lj} + I_1 B_{ik} B_{kj} - I_2 B_{ij} + I_3 \delta_{ij}$$

(applying the Cayley-Hamilton theorem to **B**)

$$= I_3 \delta_{ij} \qquad (2.6.8)$$

Combining (2.6.1), (2.6.3), (2.6.7), and (2.6.8), the components of the stress tensor become

$$T_{ij} = \frac{2}{I_3^{\frac{1}{2}}} \{ I_3 W_3 \delta_{ij} + (W_1 + I_1 W_2) B_{ij} - W_2 B_{il} B_{lj} \} \qquad (2.6.9)$$

Alternatively, this may be written in the form

$$\mathbf{T} = \alpha_0 \mathbf{I} + \alpha_1 \mathbf{B} + \alpha_2 \mathbf{B}^2 \qquad (2.6.10)$$

where the response functions

$$\alpha_0 = 2 I_3^{\frac{1}{2}} W_3, \quad \alpha_1 = \frac{2}{I_3^{\frac{1}{2}}} (W_1 + I_1 W_2), \quad \alpha_2 = -\frac{2}{I_3^{\frac{1}{2}}} W_2 \quad (2.6.11)$$

depend upon I_1, I_2, I_3.

An alternative form to (2.6.10) can be obtained by applying the Cayley–Hamilton theorem to **B**, that is,

$$\mathbf{B}^3 - I_1 \mathbf{B}^2 + I_2 \mathbf{B} - I_3 \mathbf{I} = \mathbf{0} \qquad (2.6.12)$$

Multiplying (2.6.12) by $\mathbf{B}^{-1}$ and rearranging the terms we obtain

$$\mathbf{B}^2 = I_1 \mathbf{B} - I_2 \mathbf{I} + I_3 \mathbf{B}^{-1} \qquad (2.6.13)$$

and substituting (2.6.13), (2.6.10) becomes

$$\mathbf{T} = \chi_0 \mathbf{I} + \chi_1 \mathbf{B} + \chi_{-1} \mathbf{B}^{-1} \qquad (2.6.14)$$

where

$$\chi_0 = \frac{2}{I_3^{\frac{1}{2}}}(I_2 W_2 + I_3 W_3), \quad \chi_1 = \frac{2}{I_3^{\frac{1}{2}}} W_1, \quad \chi_{-1} = -2I_3^{\frac{1}{2}} W_2$$

$$(2.6.15)$$

2.7   Cauchy elasticity

As we mentioned at the beginning of this chapter, there is a less restrictive starting point for a theory of elasticity based on (2.1.3), without the assumed existence of a strain-energy function W. To develop this approach, however, an assumption must be made concerning the behaviour of $\mathbf{T}$ under superposed rigid-body motions. (The corresponding assumption for W in the development of Green elasticity is given by (2.3.1).)

We suppose that, under the transformation (2.2.2), $\mathbf{T}$ transforms according to the rule

$$\hat{\mathbf{T}} = \mathbf{Q} \mathbf{T} \mathbf{Q}^{\mathrm{T}} \qquad (2.7.1)$$

This implies that, at each point of the body and at each instant the states of stress in the two motions differ only in orientation. If the configuration of the body in the second motion instantaneously coincides with that of the body in the first motion, then the components of the stress tensors associated with the two motions coincide. The relative motion of the two configurations therefore has no effect on the stress. This assumption again may be regarded as arising from the *principle of material frame-indifference*.

Example 2.7.1

Using (2.7.1) and arguments similar to those of Section 2.3, show that equation (2.1.3) reduces to the form

$$\mathbf{T} = \mathbf{F} \tilde{\mathbf{f}}(\mathbf{C}) \mathbf{F}^{\mathrm{T}} \qquad (2.7.2)$$

Example 2.7.2

For a material which is isotropic with respect to $\mathscr{C}$, use the arguments of Section 2.5 to show that (2.1.3) reduces to the form

$$\mathbf{T} = \bar{\mathbf{f}}(\mathbf{B}) \qquad (2.7.3)$$

and show that assumption (2.7.1) implies that the function $\bar{\mathbf{f}}$ must satisfy the identity

$$\mathbf{Q}\bar{\mathbf{f}}(\mathbf{B})\mathbf{Q}^{\mathrm{T}} = \bar{\mathbf{f}}(\mathbf{Q}\mathbf{B}\mathbf{Q}^{\mathrm{T}}) \qquad (2.7.4)$$

for all proper orthogonal tensors $\mathbf{Q}$.

A function $\bar{\mathbf{f}}$ satisfying the identity (2.7.4) is known as an isotropic tensor function, and it can be shown (using Spencer (1980) Appendix and an argument similar to that following our equation (2.6.10)) that such a function may be represented in the form

$$\mathbf{T} = \bar{\mathbf{f}}(\mathbf{B}) = f_0\mathbf{I} + f_1\mathbf{B} + f_{-1}\mathbf{B}^{-1} \qquad (2.7.5)$$

where f_0, f_1 and f_{-1} are scalar functions of the invariants of $\mathbf{B}$. Equation (2.7.5) is clearly of the same form as (2.6.14), but it should be noticed that in (2.6.14) the scalar coefficients are not independent. They are related to W through (2.6.15).

2.8   Incompressible elastic materials

In our discussion of compressible elastic materials, we postulated that the stress tensor $\mathbf{T}$ depends only on the deformation-gradient tensor $\mathbf{F}$, and that there exists a strain-energy function W, also depending only on $\mathbf{F}$, such that

$$\frac{\rho}{\rho_0}\frac{\mathrm{D}W}{\mathrm{D}t} = T_{ij}D_{ij} \qquad (2.8.1)$$

In other words, all the work done by the stresses is absorbed into W. From these assumptions, and using the results of Example 2.1.1, we were able to deduce the stress–strain relation (2.4.7).

Now for practical purposes many rubbers may be regarded as incompressible, and, as mentioned in Section 2.1, incompressible materials are capable only of isochoric deformations. The constraint

$$v_{i,i} = \operatorname{tr}\mathbf{D} = 0 \qquad (2.8.2)$$

must therefore hold, and the results of Example 2.1.1 are not applicable: the components $v_{i,j}$ are not independent. Moreover, the nature of our constitutive assumptions needs to be re-examined.

We retain the assumptions concerning the existence of the

strain-energy function which led to (2.8.1), and also the constitutive assumption which led to

$$W = W(\mathbf{C}) \tag{2.8.3}$$

However, in view of the constraint (2.8.2), we see that whatever the values of T_{ij}, the stress components $T_{ij} + p\delta_{ij}$ also satisfy (2.8.1) for arbitrary values of the scalar p. So, in place of the constitutive assumption (2.1.3) for compressible materials, we now suppose that the constitutive equation for the stress takes the form

$$T_{ij} = -p\delta_{ij} + g_{ij}(F_{kA}), \qquad \mathbf{T} = -p\mathbf{I} + \mathbf{g}(\mathbf{F}) \tag{2.8.4}$$

where $g_{ij} = g_{ji}$, and p is a scalar undetermined by the local deformation.

Example 2.8.1

Let $\mathbf{F}_0$ be an arbitrary constant tensor for which $\det \mathbf{F}_0 = 1$. Show that, in an incompressible elastic material, there exists a motion for which $\mathbf{F} = \mathbf{F}_0$ throughout the body at time $t = t_0$ for every constant value of the velocity-gradient tensor $\mathbf{L}$ for which $\operatorname{tr} \mathbf{L} = 0$. Find the body-force field required to produce such a motion.

We first note that, for any non-singular matrix $\mathbf{A}$,

$$\mathbf{A}^{-1} = J^{-1}\bar{\mathbf{A}} \tag{2.8.5}$$

where we are now denoting adj $\mathbf{A}$ by $\bar{\mathbf{A}}$, and det $\mathbf{A}$ by J. Expanding J by the first row, we have

$$J = A_{1K}\bar{A}_{K1} \tag{2.8.6}$$

Now J is a function of the elements A_{KL} of $\mathbf{A}$, and from the definition of adj $\mathbf{A}$, the element $\bar{A}_{KL}$ is independent of the corresponding element A_{KL}. Hence, from (2.8.6), we see that

$$\frac{\partial J}{\partial A_{1K}} = \bar{A}_{K1}$$

Expanding by the other rows, we may prove the general result

$$\frac{\partial J}{\partial A_{KL}} = \bar{A}_{LK} = JA_{LK}^{-1} \tag{2.8.7}$$

Consider again the motion (2.1.6), that is

$$x_i = A_{iK}(t)X_K \tag{2.8.8}$$

subject to $\mathbf{A}(t_0) = \mathbf{F}_0$, but this time with the additional constraint

$$J = \det \mathbf{A} = 1 \tag{2.8.9}$$

Then, as in Example 2.1.1, the velocity-gradient tensor $\mathbf{L}$ is given by

$$\dot{\mathbf{A}} = \mathbf{L}\mathbf{A} \tag{2.8.10}$$

However, in view of (2.8.9)

$$\dot{J} = \frac{\partial J}{\partial A_{iK}} \dot{A}_{iK} = 0 \tag{2.8.11}$$

Using (2.8.7) this gives

$$A_{Ki}^{-1} \dot{A}_{iK} = 0 \tag{2.8.12}$$

which from (2.8.10) implies $\operatorname{tr} \mathbf{L} = 0$. Conversely, $\operatorname{tr} \mathbf{L} = 0$ implies (2.8.12), which in turn implies (2.8.11), and so

$$\dot{J} = 0, \qquad J = \text{constant}$$

Hence, $J = 1$ since at $t = t_0$, $J = \det \mathbf{F}_0 = 1$.

Thus, for every constant $\mathbf{L}$ with $\operatorname{tr} \mathbf{L} = 0$ we may solve (2.8.10) for $\mathbf{A}$ as in Example 2.1.1 to obtain a motion (2.8.8) which is always possible in an incompressible material. Here again, the deformation-gradient tensor $\mathbf{F}$ depends only on t and the equation of motion may be satisfied if we take the indeterminate pressure p in (2.8.4) to be constant, and the body force as in (2.1.12).

Using (1.9.4), (2.1.1), (2.4.2), (2.8.2), and (2.8.4), equation (2.8.1) becomes

$$P_{ij}v_{i,j} = 0 \tag{2.8.13}$$

where

$$P_{ij} = g_{ij} - 2x_{i,P}x_{j,Q}\frac{\partial W}{\partial C_{PQ}} \tag{2.8.14}$$

and W is written as a symmetric function of C_{PQ}. Now P_{ij} depends only on $x_{i,K}$, but the velocity gradients $v_{i,j}$ are not independent. For example, once $v_{2,2}$ and $v_{3,3}$ are known, $v_{1,1}$ is determined by (2.8.2), and this constraint must be taken into account when discussing (2.8.13). We proceed using the method of the Lagrange multiplier. We add to the left-hand side of (2.8.13) the left-hand side of (2.8.2) multiplied by a factor $\bar{p}$, giving

$$(P_{ij} + \bar{p}\delta_{ij})v_{i,j} = 0 \qquad (2.8.15)$$

Equation (2.8.15) must hold for all motions of the body. Now we have shown in Example 2.8.1 that whatever the value of **F**, subject only to det **F** = 1, there is a motion in which the velocity gradients $v_{i,j}$ take arbitrary values except for the constraint (2.8.2).

Suppose we choose

$$\bar{p} = -P_{33} \qquad (2.8.16)$$

then $v_{3,3}$ does not appear in (2.8.15) and the remaining $v_{i,j}$ may be chosen arbitrarily while the coefficients $(P_{ij} + \bar{p}\delta_{ij})$ are fixed. Hence,

$$P_{ij} = -\bar{p}\delta_{ij} \qquad (2.8.17)$$

for i and j not both equal to 3; so, in view of (2.8.16), equation (2.8.17) holds for all i, j. Thus (2.8.4) becomes

$$T_{ij} = -(p + \bar{p})\delta_{ij} + 2x_{i,P}x_{j,Q}\frac{\partial W}{\partial C_{PQ}} \qquad (2.8.18)$$

Since p is an undetermined scalar, (2.8.18) may be written equivalently as

$$T_{ij} = -p\delta_{ij} + 2x_{i,P}x_{j,Q}\frac{\partial W}{\partial C_{PQ}} \qquad (2.8.19)$$

Equation (2.8.19) is the constitutive equation for an anisotropic, incompressible, elastic material. In contrast to equation (2.4.6), we see that the stress tensor is determined by the deformation-gradient tensor only to the extent of an arbitrary pressure p. As we shall see in the next chapter, in a particular problem p is determined by the equilibrium equations and the boundary conditions.

Example 2.8.2

By repeating the analysis given in Section 2.6 with equation (2.4.7) replaced by (2.8.19), show that the stress tensor for an incompressible, isotropic, elastic solid has the form

$$\mathbf{T} = -p\mathbf{I} + 2W_1\mathbf{B} - 2W_2\mathbf{B}^{-1} \qquad (2.8.20)$$

where p is an arbitrary scalar.

2.9   Forms of the strain-energy function

We see from (2.6.14), (2.6.15) that, once the form of the strain-energy function is known, the constitutive equation for the stress in a compressible isotropic elastic material is explicit, and from equation (2.8.20) in the incompressible case, the stress is determined to within an arbitrary pressure.

Various forms for the strain energy of compressible rubberlike solids and incompressible rubberlike solids have been proposed in the literature, and new forms are still being investigated. It is not our intention in this section to give a review of all these forms but rather to present some important examples. For further information the interested reader is referred to the articles by Blatz (1969), Ogden (1972a), and Treloar (1973). We also defer discussion of how some of the forms compare with experimental data until the next chapter.

For incompressible materials $I_3 = 1$, and

$$W = W(I_1, I_2) \tag{2.9.1}$$

Since in the reference configuration $\mathbf{C} = \mathbf{I}$, so that using (1.4.23), $I_1 = I_2 = 3$, it is convenient to regard W as a function of $I_1 - 3$ and $I_2 - 3$, and we assume that W vanishes in the reference configuration. The earliest form of W, proposed by Treloar in 1943, is

$$W = C_1(I_1 - 3) \tag{2.9.2}$$

where C_1 is a constant. This follows from a statistical theory in which vulcanised rubber is regarded as a network of long-chain molecules (for the details, see Treloar (1948)), and is often referred to as *neo-Hookean*. It appears to provide a good approximation to the behaviour of rubberlike materials (see Chapter 3), and is generally regarded as a valid prototype for this class of materials. The linear form of (2.9.1) is

$$W = C_1(I_1 - 3) + C_2(I_2 - 3) \tag{2.9.3}$$

where C_1 and C_2 are constants. This was first proposed by Mooney (1940) as the most general form admitting a linear relationship between stress and strain in simple shear, and has since been referred to as the *Mooney* form, or *Mooney–Rivlin* form. Rivlin used this form and the neo-Hookean form in much of his early work on rubber elasticity. With suitable choices of C_1 and C_2, (2.9.3) gives a marginally better fit to some of the experimental data than does (2.9.2). The terms *neo-Hookean material* and

Mooney–Rivlin material apply to materials described by the corresponding strain-energy functions.

Using (2.8.20) and the above forms of the strain-energy function, the stress tensor takes the form

$$\mathbf{T} = -p\mathbf{I} + 2C_1\mathbf{B} \qquad \text{(neo-Hookean)} \qquad (2.9.4)$$

$$\mathbf{T} = -p\mathbf{I} + 2C_1\mathbf{B} - 2C_2\mathbf{B}^{-1} \qquad \text{(Mooney–Rivlin)} \qquad (2.9.5)$$

Although both the above forms are useful and have the advantage that some problems, which would otherwise be difficult to handle, become tractable mathematically, they do not always explain the experimental data available from tests on vulcanised rubbers for all values of I_1 and I_2. Consequently, other forms have been proposed. The classic experiments of Rivlin and Saunders (see Chapter 3) suggest that, for some ranges of extension, a more accurate description of vulcanised rubber is given by the form

$$W = C_1(I_1 - 3) + f(I_2 - 3) \qquad (2.9.6)$$

where C_1 is again a constant, and f is a function for which no explicit analytical form is given, although their data implies that it is monotonic decreasing. A number of authors have considered various explicit forms for f.

A more recent development is due to Ogden (1972a). Instead of taking W as a function of I_1 and I_2, he considers it to be a function of the principal values b_1, b_2, b_3 of $\mathbf{B}$. He writes W in the form

$$W = \sum_n (\mu_n/\alpha_n)(b_1^{\alpha_n} + b_2^{\alpha_n} + b_3^{\alpha_n} - 3) \qquad (2.9.7)$$

in which the μ_n are constants, and the α_n are not necessarily integers and may be positive or negative. In common with (2.9.6), (2.9.7) does include the neo-Hookean and Mooney–Rivlin forms as special cases. The former is obtained if $n = 1$, $\alpha_1 = 2$, and the latter if $n = 2$, $\alpha_1 = 2$, $\alpha_2 = -2$. Values of the constants in (2.9.7) which currently appear to give good correlation with experimental data for vulcanised rubber are

$$\alpha_1 = 1.3, \quad \mu_1 = 6.3 \text{ kg cm}^{-2}, \quad \alpha_2 = 5.0, \quad \mu_2 = 0.012 \text{ kg cm}^{-2},$$
$$\alpha_3 = -2.0, \quad \mu_3 = -0.1 \text{ kg cm}^{-2}.$$

In reality, all rubbers are compressible although, except when subjected to very high hydrostatic pressures, vulcanised rubber undergoes only very small volume changes, so that for practical purposes it can be regarded as incompressible. For foam rubbers this is no longer the case and, motivated by the above considerations, studies have been made on the appropriate forms of W for compressible rubberlike materials, principally by Blatz and Ko (1962) and Ogden (1972b).

Using a combination of theoretical arguments and experimental results, Blatz and Ko suggested a strain-energy function of the form

$$W = \tfrac{1}{2}\mu f\left\{J_1 - 1 - \frac{1}{\nu} + \frac{1-2\nu}{\nu}J_3^{-2\nu/1-2\nu}\right\}$$

$$+ \tfrac{1}{2}\mu(1-f)\left\{J_2 - 1 - \frac{1}{\nu} + \frac{1-2\nu}{\nu}J_3^{2\nu/1-2\nu}\right\} \quad (2.9.8)$$

where μ, f, ν are constants, and

$$J_1 = I_1, \quad J_2 = I_2/I_3, \quad J_3 = I_3^{\frac{1}{2}} \quad (2.9.9)$$

We note that when $\nu = \tfrac{1}{2}$ and the material is incompressible so that $I_3 = 1$, (2.9.8) reduces to the Mooney–Rivlin form (2.9.3). In the case of 47 per cent foamed polyurethane rubber, the results of Blatz and Ko indicate that $f = 0$, $\nu = \tfrac{1}{4}$, in which case

$$W = \tfrac{1}{2}\mu\{J_2 + 2J_3 - 5\} \quad (2.9.10)$$

and clearly W is independent of I_1.

Generalising his earlier approach for incompressible materials, Ogden (1972b) proposed the form

$$W = \sum_n (\mu_n/\alpha_n)(b_1^{\alpha_n} + b_2^{\alpha_n} + b_3^{\alpha_n} - 3) + F(b_1 b_2 b_3) \quad (2.9.11)$$

in which the compressibility is accounted for by the additive function F of $b_1 b_2 b_3$. A practical assessment of appropriate forms of W for compressible materials is still awaited.

Examples 2

1. Use the relation $Q_{ik}Q_{jk} = \delta_{ij}$ (see (2.2.3)), for time-dependent orthogonal tensors $\mathbf{Q}$, to show that the tensor $\mathbf{\Omega}$ with components

$\Omega_{ij} = \dot{Q}_{ik}Q_{jk}$ is skew-symmetric, that is, $\Omega_{ij} = -\Omega_{ji}$, where the superposed dot stands for differentiation with respect to time.

2. Use the relation (2.2.2) to show that

$$\frac{\partial \hat{v}_i}{\partial \hat{x}_j} = Q_{ip}Q_{jq}\frac{\partial v_p}{\partial x_q} + \Omega_{ij}$$

and deduce that the rate-of-deformation tensor **D**, and spin tensor **W**, transform according to the rules

$$\hat{\mathbf{D}} = \mathbf{QDQ}^{\mathrm{T}}, \qquad \hat{\mathbf{W}} = \mathbf{QWQ}^{\mathrm{T}} + \mathbf{\Omega}$$

3. The constitutive equation for a certain material is postulated in the form $\mathbf{T} = \mathbf{f}(\mathbf{D}, \mathbf{W})$. Show that (2.7.1) can be satisfied if and only if the constitutive equation reduces to the form $\mathbf{T} = \bar{\mathbf{f}}(\mathbf{D})$.

4. By integrating the equations

$$\frac{\partial W}{\partial J_1} = \tfrac{1}{2}\mu f, \quad \frac{\partial W}{\partial J_2} = \tfrac{1}{2}\mu(f-1), \quad \frac{\partial W}{\partial J_3} = \mu\{(1-f)J_3^{(4\nu-1)/(1-2\nu)} - fJ_3^{-1/(1-2\nu)}\}$$

where μ, f, and ν are constants, derive (2.9.8).

Exact solutions

Combining the constitutive theory developed in the previous chapter with the field equations (1.9.4), (1.11.14), and (1.11.18), we have a non-linear theory for a homogeneous, isotropic, elastic material. In this chapter we investigate the possibility of finding exact solutions of the equations without restriction on the form of the strain-energy function except for that imposed in some cases by the incompressibility condition.

3.1 Basic equations. Boundary conditions

Since (2.6.14) implies the symmetry of the stress tensor, equation (1.11.18) is satisfied identically. Restricting our attention to bodies maintained in equilibrium, the remaining equations which have to be satisfied for a compressible material are

$$\rho J = \rho_0 \qquad (3.1.1)$$

and

$$T_{ij,j} + \rho b_i = 0 \qquad (3.1.2)$$

where

$$T_{ij} = \chi_0 \delta_{ij} + \chi_1 B_{ij} + \chi_{-1} B_{ij}^{-1} \qquad (3.1.3)$$

and χ_0, χ_1, χ_{-1} are functions of I_1, I_2, and I_3.

One way to maintain a body in equilibrium is to apply suitable surface tractions on its boundary. This type of problem gives rise to *traction boundary conditions* involving the stress vector. Let us consider a few typical examples. When a body of arbitrary shape is held in equilibrium under the action of a hydrostatic pressure $P(>0)$ per unit area of the surface of the deformed configuration as shown in Fig. 3.1,

$$\mathbf{t}(\mathbf{n}) = -P\mathbf{n} \qquad (3.1.4)$$

As another example, consider a block extended in the 1-direction and maintained in equilibrium by applying a uniform

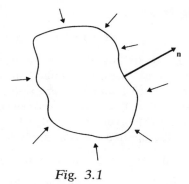

Fig. 3.1

tension T per unit area of the deformed configuration on its end faces as shown in Fig. 3.2. Suppose that after the deformation the block occupies the region

$$-a \leqslant x_1 \leqslant a, \qquad -b \leqslant x_2 \leqslant b, \qquad -c \leqslant x_3 \leqslant c$$

The end faces $x_1 = \pm a$ are both perpendicular to the 1-direction but they have different outward unit normals. On the face $x_1 = -a$ the outward unit normal is $-\mathbf{e}_1$ so that our boundary condition specifies $\mathbf{t}(-\mathbf{e}_1)$ and requires that

$$\mathbf{t}(-\mathbf{e}_1) = -T\mathbf{e}_1 \tag{3.1.5}$$

But at any point on this face, by (1.11.7),
$\mathbf{t}(-\mathbf{e}_1) = -\mathbf{t}(\mathbf{e}_1) = -(T_{11}, T_{12}, T_{13})$, so that an equivalent statement is

$$T_{11} = T, \qquad T_{12} = T_{13} = 0 \tag{3.1.6}$$

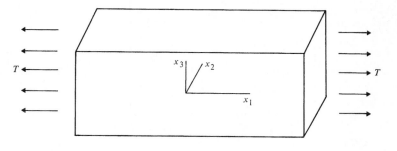

Fig. 3.2

Likewise, on the face $x_1 = a$,

$$\mathbf{t}(\mathbf{e}_1) = T\mathbf{e}_1 \qquad (3.1.7)$$

which again gives rise to (3.1.6). Since we are not applying any forces to the remaining faces, the applied surface traction on these faces is zero. Such boundaries are said to be *traction-free*. We use this term rather than 'stress-free' since in many cases the surface is not stress-free even though it is free of applied traction. The vanishing of the applied traction implies

$$\mathbf{t}(\mathbf{e}_2) = \mathbf{0} \quad \text{on } x_2 = b, \quad \mathbf{t}(-\mathbf{e}_2) = \mathbf{0} \quad \text{on } x_2 = -b \qquad (3.1.8)$$

and so the stresses T_{21}, T_{22}, T_{23} are zero on $x_2 = \pm b$, but in these faces $T_{11} \neq 0$. Similarly, the stresses T_{31}, T_{32}, T_{33} are zero on $x_3 = \pm c$ but again $T_{11} \neq 0$.

Consider next a hollow circular cylinder as shown in Fig. 3.3, which is held in equilibrium under suitable surface tractions. In the deformed configuration let the inner and outer radii be a_1 and a_2 respectively. The surface $r = a_1$ is subjected to a uniform

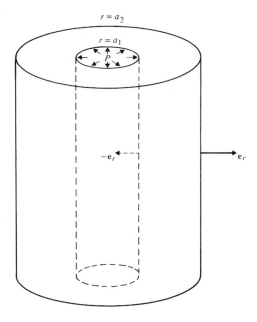

Fig. 3.3

pressure P and the outer surface $r = a_2$ is traction-free. Tractions are also applied to the end faces. On the surface $r = a_1$ the outward unit normal is $-\mathbf{e}_r$ so that our boundary condition on this surface specifies $\mathbf{t}(-\mathbf{e}_r)$ and requires that

$$\mathbf{t}(-\mathbf{e}_r) = P\mathbf{e}_r \tag{3.1.9}$$

on $r = a_1$. The outer surface is traction-free provided

$$\mathbf{t}(\mathbf{e}_r) = \mathbf{0} \tag{3.1.10}$$

on $r = a_2$. Now $t_i = T_{ij}n_j$ and $\mathbf{e}_r = r^{-1}(x_1, x_2, 0)$, so that on $r = a_2$,

$$t_i(\mathbf{e}_r) = T_{i\alpha}x_\alpha/a_2, \qquad \alpha = 1, 2 \tag{3.1.11}$$

Thus a statement equivalent to (3.1.10) is

$$t_1 = T_{11}\frac{x_1}{a_2} + T_{12}\frac{x_2}{a_2} = 0, \quad t_2 = T_{21}\frac{x_1}{a_2} + T_{22}\frac{x_2}{a_2} = 0,$$
$$t_3 = T_{31}\frac{x_1}{a_2} + T_{32}\frac{x_2}{a_2} = 0 \tag{3.1.12}$$

The relation (3.1.9) may be expanded similarly.

We see from the above examples that in a particular problem we are required to solve (3.1.1) and (3.1.2) subject to prescribed boundary conditions. In writing down these conditions it is important to be able to identify the outward unit normal to the surface under consideration and also to realise which components of the stress tensor are being specified by the applied surface tractions. In the above examples the body is considered in the deformed configuration and T and P are measured per unit area in this state. It is therefore appropriate to use the Cauchy stress tensor $\mathbf{T}$. In some situations the applied forces may be measured per unit area of the reference configuration, in which case the Piola–Kirchhoff stress tensors are more useful.

For an incompressible material there is no volume change, so that we have the condition

$$J = 1 \tag{3.1.13}$$

and the density has the value ρ_0 for all time. Equation (3.1.1) therefore reduces to an identity, and the equations which have to be satisfied in this case are (3.1.13) and (3.1.2), where now from (2.8.20),

$$T_{ij} = -p\delta_{ij} + 2W_1 B_{ij} - 2W_2 B_{ij}^{-1} \tag{3.1.14}$$

p being an unknown scalar. As we shall see later, in a particular problem p is determined by the equilibrium equations (3.1.2) and the specified boundary conditions.

3.2   Inverse method

The complexity of the governing equations, (3.1.1) to (3.1.3), or (3.1.2), (3.1.13), and (3.1.14), usually prevents us from solving them generally. Instead of seeking general solutions and then subjecting them to given boundary conditions, we employ what is known as the *inverse method*. We assume a suitable form for the deformation; we use the constitutive equations to find the associated stresses, and then consider what restrictions, if any, are imposed on the deformation by the equilibrium equation. We then deduce the surface tractions necessary to maintain the deformation. In this way we build up a set of exact solutions to a number of boundary-value problems, some of which are of considerable importance experimentally. We proceed now to examine the homogeneous and non-homogeneous deformations, described in Chapter 1, as a basis for this inverse method.

3.3   Homogeneous deformations

For a homogeneous deformation the deformation-gradient tensor $\mathbf{F}$ is constant (see (1.5.1)) and so, since $\mathbf{T} = \mathbf{f}(\mathbf{F})$ (see (2.1.3)), the stress tensor $\mathbf{T}$ is also constant throughout a compressible material. It follows that the equilibrium equations (3.1.2) are satisfied only when the body force $\mathbf{b}$ is zero. In other words, these deformations may be produced by surface tractions alone, regardless of the material. This result is important physically since it is relatively easy to apply forces to a boundary. For incompressible materials we have seen that the stress is determined by the deformation gradient only to within a hydrostatic pressure p. However, if p is constant the above result also applies to incompressible bodies.

   In order to see what type of boundary conditions would produce a particular deformation, and also to consider observable effects which may be of value in experiments, we consider the deformations introduced in Chapter 1. Throughout we assume that body force is absent, so that for compressible materials the equilibrium equations are identically satisfied.

3.4 Pure homogeneous deformation of a compressible material

For the deformation (see Example 1.5.1)

$$x_1 = \lambda_1 X_1, \qquad x_2 = \lambda_2 X_2, \qquad x_3 = \lambda_3 X_3 \qquad (3.4.1)$$

where λ_1, λ_2, λ_3 are non-zero constants, $\mathbf{B}$ and $\mathbf{B}^{-1}$ are given by

$$\mathbf{B} = \begin{pmatrix} \lambda_1^2 & 0 & 0 \\ 0 & \lambda_2^2 & 0 \\ 0 & 0 & \lambda_3^2 \end{pmatrix} \qquad \mathbf{B}^{-1} = \begin{pmatrix} \lambda_1^{-2} & 0 & 0 \\ 0 & \lambda_2^{-2} & 0 \\ 0 & 0 & \lambda_3^{-2} \end{pmatrix} \qquad (3.4.2)$$

and the principal invariants are given by (1.5.11). The response functions appearing in (3.1.3) are therefore constant, and the stress components are

$$T_{11} = \chi_0 + \chi_1 \lambda_1^2 + \chi_{-1} \lambda_1^{-2}, \quad T_{22} = \chi_0 + \chi_1 \lambda_2^2 + \chi_{-1} \lambda_2^{-2},$$
$$T_{33} = \chi_0 + \chi_1 \lambda_3^2 + \chi_{-1} \lambda_3^{-2}, \quad T_{ij} = 0 \quad (i \neq j) \qquad (3.4.3)$$

where

$$\chi_i = \chi_i (\lambda_1^2 + \lambda_2^2 + \lambda_3^2, \ \lambda_1^2 \lambda_2^2 + \lambda_2^2 \lambda_3^2 + \lambda_3^2 \lambda_1^2, \ \lambda_1^2 \lambda_2^2 \lambda_3^2)$$

$i = -1, 0, 1$. We see therefore that only normal stresses are present on surfaces parallel to the coordinate planes.

(i) Dilatation

The deformation (3.4.1) in the special case when $\lambda_1 = \lambda_2 = \lambda_3 = \alpha$ (with $\alpha > 0$, to satisfy $J > 0$) is known as a uniform dilatation. From (3.4.3) the stress components are

$$T_{ij} = -P(\alpha^2)\delta_{ij} \qquad (3.4.4)$$

which correspond to a hydrostatic pressure

$$P(\alpha^2) = -\chi_0 - \chi_1 \alpha^2 - \chi_{-1} \alpha^{-2} \qquad (3.4.5)$$

where χ_0, χ_1, χ_{-1} depend only on α^2. Since $t_i = T_{ij} n_j$, we see that the stress vector $\mathbf{t}$ is of the form

$$\mathbf{t} = -P\mathbf{n} \qquad (3.4.6)$$

so that for this deformation to be maintained, the stress vector must be normal to the surface at each point of the boundary. As we have seen earlier, a boundary condition of this type arises when a body is subjected to a uniform pressure $(P > 0)$, or tension $(P < 0)$, on the boundary.

One would expect that the volume of a compressible material held in equilibrium under the action of a uniform pressure should be less than its volume before deformation and that, under the action of a uniform tension, the volume should be greater than its initial volume. Since $J = \alpha^3$ and $dV = J\,dV_0$, an equivalent statement is that

$$\alpha < 1 \quad \text{when} \quad P > 0, \qquad \alpha > 1 \quad \text{when} \quad P < 0 \qquad (3.4.7)$$

Also if no traction is applied on the boundary, the volume is expected to remain unchanged, so that

$$\alpha = 1 \quad \text{when} \quad P = 0 \qquad (3.4.8)$$

Further, when the applied pressure is increased the volume should decrease, and vice versa. Likewise, when the applied tension is increased the volume should increase. In other words, P should be a monotonic decreasing function of α, in which case

$$dP/d\alpha < 0 \qquad (3.4.9)$$

Of course, (3.4.9) and (3.4.8) imply (3.4.7). In view of (3.4.5), (3.4.9) places some restrictions on the response functions χ_0, χ_1, χ_{-1} and hence, using (2.6.15), on the form of the strain-energy function.

(ii) Simple extension

When $\lambda_1 = \alpha$ and $\lambda_2 = \lambda_3 = \beta$, the deformation (3.4.1) represents a uniform extension or contraction in the 1-direction together with equal extensions or contractions in the lateral 2- and 3-directions. From (3.4.3) we see that in this deformation $T_{22} = T_{33}$. The simplest stress system arises when, if possible, $T_{22} = T_{33} = 0$ throughout the body, in which case we require

$$\chi_0 + \chi_1 \beta^2 + \chi_{-1} \beta^{-2} = 0 \qquad (3.4.10)$$

where χ_0, χ_1, χ_{-1} depend only on α^2 and β^2. In this case the only non-zero stress is T_{11}. Since T_{11} is constant it has the same value within the body and on the boundary. Putting $T_{11} = T$, this stress system would certainly satisfy the boundary conditions discussed earlier (Section 3.1) for a block maintained in equilibrium under the action of a uniform tension T on its end faces, the other faces being traction-free.

However, for a given extension α it is not obvious that (3.4.10) has a single positive root β^2. If (3.4.10) has no root then uniform extension cannot be effected by applying a tension T on the end faces; other surface tractions are necessary. If (3.4.10) has more than one root, and χ_0, χ_1, and χ_{-1} are single-valued functions, then

$$T_{11} = \chi_0 + \chi_1 \alpha^2 + \chi_{-1} \alpha^{-2} \tag{3.4.11}$$

has a different value for different values of β^2. That is, there are more than one tensile stresses which produce a given extension with the remaining faces traction-free. Each value of β would correspond to a different displacement in the 2- and 3-directions. In other words, by applying different tensions on the boundary, the same extension is still produced. If (3.4.10) may be solved uniquely for β^2 in terms of α^2, we may substitute this value of β^2 into (3.4.11) which then takes the form

$$T_{11} = T(\alpha^2) \tag{3.4.12}$$

When we apply a tension in the 1-direction with the other stresses being zero, we expect the specimen to increase in length in this direction, whereas when we apply a pressure the length should decrease. Thus

$$T(\alpha^2) > 0 \quad \text{when} \quad \alpha > 1, \qquad T(\alpha^2) < 0 \quad \text{when} \quad \alpha < 1 \tag{3.4.13}$$

Also, if no tension is applied on the boundary the length should remain unchanged, so that when $T(\alpha^2) = 0$, $\alpha = 1$. Further, when the applied tension is increased the extension should increase, and vice versa. Hence T should be a monotonically increasing function of α, and so

$$dT/d\alpha > 0 \tag{3.4.14}$$

Since $T = 0$ when $\alpha = 1$, (3.4.14) implies (3.4.13). Clearly from (3.4.11) and (3.4.12) this inequality places a further restriction on the response functions and hence on the form of the strain-energy function.

3.5   Pure homogeneous deformation of an incompressible material

For the deformation (3.4.1) to be possible in an incompressible material, it must satisfy the constraint $J = 1$. This gives

$$\lambda_1\lambda_2\lambda_3 = 1 \tag{3.5.1}$$

so that, in contrast to the compressible case, only two of the constants λ_1, λ_2, λ_3 can be chosen arbitrarily; the third is determined by (3.5.1). Also, for the equations of equilibrium (3.1.2) to be satisfied in the absence of body force, the pressure p appearing in (3.1.14) must satisfy $p_{,i} = 0$, giving

$$p = p_0 = \text{constant} \tag{3.5.2}$$

The stress system which now follows from (3.1.14) using (3.4.2), (3.5.1), and (3.5.2) is

$$T_{11} = -p_0 + 2W_1\lambda_1^2 - 2W_2\lambda_1^{-2}, \quad T_{22} = -p_0 + 2W_1\lambda_2^2 - 2W_2\lambda_2^{-2}, \tag{3.5.3}$$

$$T_{33} = -p_0 + 2W_1\lambda_1^{-2}\lambda_2^{-2} - 2W_2\lambda_1^2\lambda_2^2, \quad T_{ij} = 0 \quad (i \neq j)$$

where the principal invariants are given by

$$I_1 = \lambda_1^2 + \lambda_2^2 + \lambda_1^{-2}\lambda_2^{-2}, \qquad I_2 = \lambda_1^2\lambda_2^2 + \lambda_1^{-2} + \lambda_2^{-2} \tag{3.5.4}$$

Although this stress system is similar to that of (3.4.3) in the sense that only normal stresses are present on surfaces parallel to the coordinate planes, it differs from (3.4.3) in that they now contain an arbitrary constant pressure p_0. The appearance of this term is one of the reasons why incompressible materials are easier to deal with mathematically than compressible ones. Before considering the general situation in more detail, let us return to the simpler deformations discussed in the previous section.

(i) Dilatation

In this case (3.5.1) requires that $\alpha = 1$, so that there is no deformation. We would expect this on physical grounds since an incompressible material cannot change its volume.

(ii) Simple extension

The lateral contraction (or extension) β is now given by (3.5.1) in terms of α. The boundary conditions $T_{22} = T_{33} = 0$, appropriate to the block subject to a tension T are satisfied if

$$p_0 = 2\alpha^{-1}W_1 - 2\alpha W_2 \qquad (3.5.5)$$

This result is very different from its counterpart in a compressible material since by taking p_0 as given in (3.5.5) we can always satisfy the boundary conditions $T_{22} = T_{33} = 0$, and effect simple extension by applying a uniaxial tension T_{11} given by

$$T_{11} = 2(\alpha^2 - \alpha^{-1})W_1 + 2(\alpha - \alpha^{-2})W_2 \qquad (3.5.6)$$

(iii) Stretching of a plane sheet

Returning to the general deformation, we see from (3.5.3) that by suitable choice of p_0 any one of the stresses can be made to vanish. As a particular example, consider the stretching of a plane rectangular sheet of incompressible material by forces in its plane. Suppose that the axes are chosen with the major faces perpendicular to the 3-direction (see Fig. 3.4) and free of traction, so that $\mathbf{t}(\mathbf{e}_3) = \mathbf{0}$. Since for this deformation $T_{31} = T_{32} = 0$ throughout the body, and in particular on the boundary, this condition is

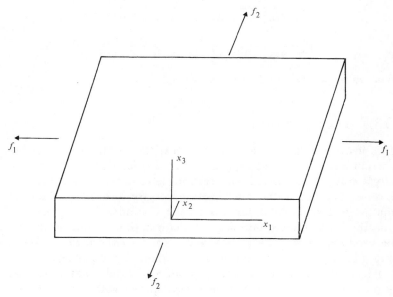

Fig. 3.4

satisfied if $T_{33} = 0$. This is the case if

$$p_0 = 2W_1\lambda_1^{-2}\lambda_2^{-2} - 2W_2\lambda_1^2\lambda_2^2 \qquad (3.5.7)$$

and the non-zero stress components are

$$\begin{aligned}
T_{11} &= 2(\lambda_1^2 - \lambda_1^{-2}\lambda_2^{-2})(W_1 + \lambda_2^2 W_2) \\
T_{22} &= 2(\lambda_2^2 - \lambda_1^{-2}\lambda_2^{-2})(W_1 + \lambda_1^2 W_2)
\end{aligned} \qquad (3.5.8)$$

These are the forces, measured per unit area of the deformed configuration, required to maintain the deformation. However, T_{11} and T_{22} may be expressed in terms of the forces f_1, f_2 per unit length of edge measured in the undeformed state, since

$$f_1 = hT_{11}/\lambda_1, \qquad f_2 = hT_{22}/\lambda_2 \qquad (3.5.9)$$

where h is the thickness of the sheet in its undeformed state.

Example 3.5.1

Derive the relations (3.5.9).

Solving (3.5.8) for W_1 and W_2, we obtain

$$\begin{aligned}
W_1 &= \frac{1}{2(\lambda_1^2 - \lambda_2^2)}\left\{ \frac{\lambda_1^2 T_{11}}{\lambda_1^2 - \lambda_1^{-2}\lambda_2^{-2}} - \frac{\lambda_2^2 T_{22}}{\lambda_2^2 - \lambda_1^{-2}\lambda_2^{-2}} \right\} \\
W_2 &= \frac{1}{2(\lambda_2^2 - \lambda_1^2)}\left\{ \frac{T_{11}}{\lambda_1^2 - \lambda_1^{-2}\lambda_2^{-2}} - \frac{T_{22}}{\lambda_2^2 - \lambda_1^{-2}\lambda_2^{-2}} \right\}
\end{aligned} \qquad (3.5.10)$$

The values of I_1 and I_2 are given in (3.5.4).

3.6   Experiments

It is instructive at this stage to see how the results of the previous section have been used experimentally. So far we have found the stress systems associated with certain deformations, and these have enabled us to find the forces necessary to produce the deformation in terms of the derivatives of the strain-energy function and the parameters describing the deformation. These results can be used experimentally in two ways. Knowing the form of the strain energy as a function of the invariants for a particular material, the forces necessary to produce the deformation can be explicitly calculated and the results compared with experiment. Alternatively, the results of the experiments can be used to give some information about the form of the strain-energy function for the particular material being used. Much of the experimental work has been done for vulcanised rubber which is assumed to be incompressible. We restrict our attention here to this case, and describe below (and in later sections) the experiments performed by Treloar, and Rivlin and Saunders. The results obtained in their experiments are still used in current work when testing new theoretical forms for W. Experiments on compressible foam rubbers are described in the article by Blatz (cited in Section 2.9).

(i) The experiments of Treloar

During the 1940s Treloar designed a series of experiments using vulcanised natural rubber to test the applicability of the neo-Hookean form of the strain-energy function. We mention two of these experiments to illustrate how the results derived in the previous section can be used to test a given form of strain energy.

One of the easiest tests to perform is simple extension. For the neo-Hookean form (2.9.2)

$$W_1 = C_1, \qquad W_2 = 0 \qquad (3.6.1)$$

so that, using (3.5.6),

$$T_{11} = 2C_1(\alpha^2 - \alpha^{-1}) \qquad (3.6.2)$$

where T_{11} is the force measured per unit area of the deformed configuration. The force T per unit area of the undeformed configuration, required to produce the extension is

$$T = T_{11}/\alpha = 2C_1(\alpha - \alpha^{-2}) \qquad (3.6.3)$$

By increasing the weights on the specimen at intervals of about one minute, Treloar obtained the stress–strain curve given in Fig. 3.5. It is seen that beyond about 50 per cent extension ($\alpha = 1.5$) the experimental curve falls below the theoretical one and it rises rapidly for higher extension. Similar results were obtained in other experimental situations indicating that for these deformations the neo-Hookean form is a reasonable approximation for small extensions only.

A more general type of deformation is the pure homogeneous deformation (3.4.1). This can be produced experimentally by stretching a thin square sheet of rubber. In Treloar's experiment the square test-piece had five projecting lugs on each side as

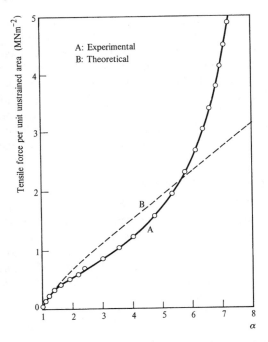

Fig. 3.5   From Treloar, 1973. © The Institute of Physics

indicated schematically in Fig. 3.6, and strings were tied to these for the application of loads; the surface of the sheet was marked out with lines forming a square lattice. With the sheet placed horizontally, the three middle lugs on one side were loaded by means of three equal weights attached to strings passing over pulleys, while the strings attached to the two outermost lugs were secured to a rectangular frame. Similar arrangements were used on the other three sides, and by applying different weights, different extensions were obtained. The appearance of the stretched sheet is shown in Fig. 3.7. By measuring the lengths of the sides of the rectangles, drawn on the sheet, in their deformed and undeformed states, one can find λ_1 and λ_2. In the experiment it was assumed that the forces operative in producing the pure homogeneous deformation of the rectangle ABCD were the loads applied to the three central lugs on each side of the test piece. The forces f_1, f_2 were then known, and the values of T_{11} and T_{22} were obtained from (3.5.9).

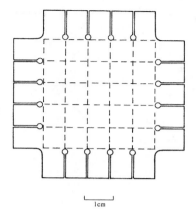

Fig. 3.6 From Treloar, 1958

For a neo-Hookean material, (3.5.8) and (2.9.2) give

$$T_{11} = 2C_1(\lambda_1^2 - \lambda_1^{-2}\lambda_2^{-2}), \qquad T_{22} = 2C_1(\lambda_2^2 - \lambda_1^{-2}\lambda_2^{-2}) \quad (3.6.4)$$

so that, on plotting T_{11} against $\lambda_1^2 - \lambda_1^{-2}\lambda_2^{-2}$, a straight line passing through the origin should result. This was found not to be the case. When a Mooney–Rivlin form of strain energy is taken, using (2.9.3) and (3.5.8) we find

$$\begin{aligned} T_{11} &= 2(\lambda_1^2 - \lambda_1^{-2}\lambda_2^{-2})(C_1 + \lambda_2^2 C_2) \\ T_{22} &= 2(\lambda_2^2 - \lambda_1^{-2}\lambda_2^{-2})(C_1 + \lambda_1^2 C_2) \end{aligned} \qquad (3.6.5)$$

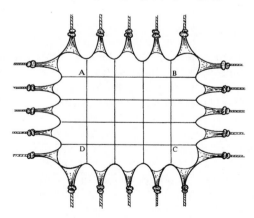

Fig. 3.7 From Treloar, 1958

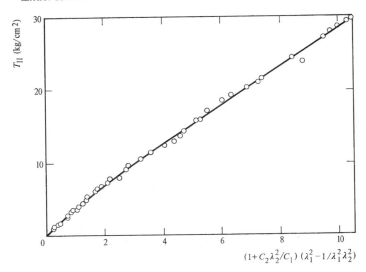

$$(1+C_2\lambda_2^2/C_1)\,(\lambda_1^2-1/\lambda_1^2\lambda_2^2)$$

Fig. 3.8   From Treloar, 1958

giving a linear relationship between T_{11} and
$(\lambda_1^2-\lambda_1^{-2}\lambda_2^{-2})(1+C_2\lambda_2^2/C_1)$. The relationship obtained with the
value $C_2/C_1 = 0.05$ is shown in Fig. 3.8. Although the deviations
from linearity are less pronounced than in the case of the neo-Hookean
form, they are not completely eliminated. Thus, while giving a better fit
than the neo-Hookean form, the Mooney–Rivlin form is still not
entirely adequate.

(ii) The experiments of Rivlin and Saunders

As mentioned earlier, an alternative use of the theoretical results
derived in Section 3.5 is to determine the form of the strain-
energy function for a given material. Not all deformations are
suitable for this purpose. For example, in simple extension only
one parameter α arises, and it is impossible to vary I_1 and I_2
independently. A more fruitful approach is provided by the deformation discussed in Section 3.5 (iii). However, Treloar's experiment is not easily adapted to determine with any precision the
dependence of W on I_1 and I_2. Rivlin and Saunders (1951) used
the arrangement shown in Fig. 3.9. One difference between their
arrangement and that of Treloar's is that springs were used instead of weights, so that rapid, continuous variation of the applied

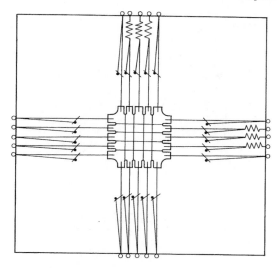

Fig. 3.9   From Rivlin and Saunders, 1951

stress was possible. Again it was assumed that the forces operative in producing the pure homogeneous deformation of the central nine squares were those applied to the central three strings
on each side of the test-piece, and the forces were taken as the
means of the tensions measured in the central three strings on the
appropriate side of the test-piece.

If, as λ_1 is varied, λ_2 is also varied in such a manner that I_2
remains constant, then using (3.5.10) we obtain values of W_1 and
W_2 for various values of I_1 at constant I_2. Similarly, variations of
λ_1 and λ_2 such that I_1 remains constant while I_2 varies give the
dependence of W_1 and W_2 on I_2 for constant values of I_1. From
(3.5.4) we see that for a fixed value of I_1 or I_2, λ_1 and λ_2 cannot
be varied arbitrarily; we must have

$$\lambda_2^2 = \tfrac{1}{2}\{I_1 - \lambda_1^2 \pm [(I_1 - \lambda_1^2)^2 - 4\lambda_1^{-2}]^{\frac{1}{2}}\}$$
$$\lambda_2^2 = \tfrac{1}{2}\lambda_1^{-2}\{I_2 - \lambda_1^{-2} \pm [(I_2 - \lambda_1^{-2})^2 - 4\lambda_1^2]^{\frac{1}{2}}\} \tag{3.6.6}$$

which shows that the variation of λ_2 with λ_1 for fixed values of I_1
or I_2 is restricted to some closed curves in a $\lambda_1\lambda_2$-plane. These
are shown in Fig. 3.10 for some constant values of I_1 and I_2, the
broken curves representing the line

$$\lambda_2 = \lambda_1^{-\frac{1}{2}} (f_2 = 0), \qquad \lambda_1 = \lambda_2^{-\frac{1}{2}} (f_1 = 0) \tag{3.6.7}$$

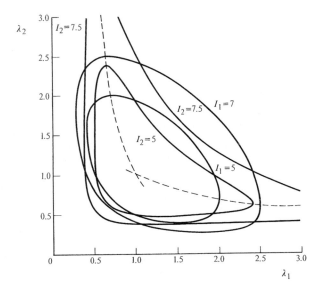

Fig. 3.10   From Rivlin and Saunders, 1951

which correspond to simple extensions parallel to the 1- and 2-directions respectively.

The experimental results of Rivlin and Saunders obtained by the above method are shown in Fig. 3.11 and suggest a form of strain-energy function given by

$$W = C_1(I_1 - 3) + f(I_2 - 3) \qquad (3.6.8)$$

in which W_1 is the constant C_1, and $W_2 = df/dI_2$ is a function of I_2 only. The ratio W_2/W_1 has an approximate value of $\frac{1}{8}$ for low values of I_2, and decreases steadily with increasing I_2. In this experiment the range of values for I_1 and I_2 is $5 < I_1 < 12$ and $5 < I_2 < 30$. There is some scatter of the experimental readings and, since the calculations based on (3.5.10) are sensitive to quite small experimental errors, it is possible that the experimental results could equally well be represented by some slightly different dependence of W_1 and W_2 on I_1 and I_2. Nevertheless, the form (3.6.8) for W enables the theory to predict results for other types of deformation which are in good agreement with experiment.

In calculating W_1 and W_2 it is seen from (3.5.10) that any error in T_{11} or T_{22} is reflected in a greater percentage error in

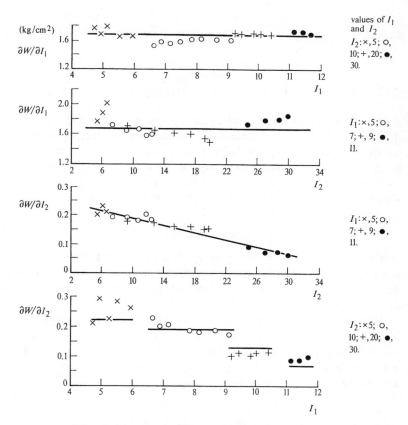

Fig. 3.11   From Rivlin and Saunders, 1951

the calculation of W_1 and W_2. This magnification of experimental error becomes more and more pronounced as λ_1 and λ_2 approach the same value. In the case when $\lambda_1 = \lambda_2$, T_{11} should be equal to T_{22}, since we are dealing with isotropic materials, so that the values of W_1 and W_2 become indeterminate. Consequently, this sets a lower limit for which the method described above can be used to determine the derivatives of W. In practice, the method was not found convenient for values of I_1 and I_2 less than 5.

For lower values of I_1 and I_2, Rivlin and Saunders gained some indication of the dependence of W_1 and W_2 on I_1 and I_2 by carrying out further experiments. To within experimental error they found that the form (3.6.8) was also reasonable for the lower values of I_1 and I_2. It is worth noting that the results of these experiments cannot be explained using the Mooney–Rivlin form of the strain energy.

3.7   Simple shear of a compressible material

Although simple shear would appear to be a deformation which is difficult to produce experimentally because of the surface tractions needed to maintain it (see below), it is probably the simplest example which illustrates that large deformations are different from infinitesimal deformations (Chapters 4–7) not only in magnitude but also in the novel effects they produce.

For such a deformation (see (1.5.12)),

$$x_1 = X_1 + \kappa X_2, \quad x_2 = X_2, \quad x_3 = X_3 \qquad (3.7.1)$$

and we consider the shearing by applied surface tractions of a cube with faces initially parallel to the coordinate planes. A typical cross-section in a plane of shear of the initial and final configurations is shown in Fig. 1.3.

Now **B** is given by (1.5.14), $\mathbf{B}^{-1}$ by (1.5.16) and the principal invariants by (1.5.17). In particular, $I_3 = 1$, and so we have an example of an isochoric deformation which can take place in all compressible materials. Using (3.1.3) the stress components associated with this deformation are

$$
\begin{aligned}
T_{11} &= \chi_0 + (1+\kappa^2)\chi_1 + \chi_{-1}, \quad T_{22} = \chi_0 + \chi_1 + (1+\kappa^2)\chi_{-1} \\
T_{33} &= \chi_0 + \chi_1 + \chi_{-1}, \quad T_{12} = \kappa(\chi_1 - \chi_{-1}), \quad T_{23} = T_{31} = 0
\end{aligned}
\qquad (3.7.2)
$$

where $\chi_i = \chi_i(3+\kappa^2, 3+\kappa^2, 1)$, $i = -1, 0, 1$. Thus we see that both normal and shear stresses are present on surfaces parallel to the coordinate planes. As in the previous deformation, the stress components are constant.

To consider the relation between the shear stress and the angle of shear, we write

$$\mu(\kappa^2) = \chi_1(3+\kappa^2, 3+\kappa^2, 1) - \chi_{-1}(3+\kappa^2, 3+\kappa^2, 1) \qquad (3.7.3)$$

then

$$T_{12} = \mu(\kappa^2)\kappa \qquad (3.7.4)$$

and $\mu(\kappa^2)$ is called the *generalised shear modulus*. From (3.7.4) we see that T_{12} is an odd function of κ, so that if we reverse the direction of shear we alter the sign of the shear stress. Physically we would expect the shear stress acting on a surface with normal in the 2-direction to be in the direction in which the surface has been displaced, so that we expect

$$\mu(\kappa^2) > 0 \qquad (3.7.5)$$

This follows from (3.7.3) if, in particular,

$$\chi_1 > 0, \qquad \chi_{-1} < 0 \qquad (3.7.6)$$

These inequalities are found to hold in the case of incompressible vulcanised rubber.

In the case of infinitesimal deformations (Section 4.7) simple shear can be maintained by applying only shear stresses on the faces of the specimen, and so a natural question to ask is whether, in a finite deformation, the normal stresses can all vanish with $\kappa \neq 0$. Since the stresses have constant values, this would be the case if

$$\chi_0 + \chi_1 + \chi_{-1} = 0$$
$$\chi_0 + \chi_1 + (1 + \kappa^2)\chi_{-1} = 0 \qquad (3.7.7)$$
$$\chi_0 + (1 + \kappa^2)\chi_1 + \chi_{-1} = 0$$

for $\kappa \neq 0$, from which it follows that $\chi_0 = \chi_1 = \chi_{-1} = 0$. In this case T_{12} is also zero. These conditions are therefore satisfied only in a highly degenerate material in which a simple shear can be produced in the absence of all stress. We therefore conclude that, for all materials exhibiting physically reasonable response, the simple shear (3.7.1) cannot be produced by applying only shear stresses on surfaces parallel to the coordinate planes; normal stresses are also necessary. In fact, if such normal stresses are not applied, the material will tend to contract or expand. This result was apparently conjectured by Kelvin and is often called the *Kelvin effect*. The above argument is independent of the form of W, and so the conclusions hold for all isotropic elastic materials. In a particular material for which W is specified, two of the normal stresses may vanish but, except in the case of the degenerate material mentioned above, never all three.

From (3.7.2) we see that

$$T_{11} - T_{22} = \kappa T_{12} \qquad (3.7.8)$$

This relation does not depend on the response functions χ_0, χ_1,

χ_{-1} and so holds for all isotropic elastic materials; such a result is called a *universal relation*. If one finds experimentally that (3.7.8) is not satisfied then one may conclude that the material under investigation is not an isotropic elastic material. One consequence of (3.7.8) is that, except in the case of the degenerate material mentioned above, T_{11} cannot be equal to T_{22}. Unless these unequal normal stresses are applied, the specimen may be expected to extend or contract in the 1- and 2-directions. Using a special theory of elasticity, Poynting (1909) noticed a similar phenomenon and performed a series of torsion experiments to illustrate the lengthening of a wire if no normal force was applied. The existence of unequal normal stresses is often referred to as the *Poynting effect*.

We now consider the normal stress N and the shear stress S which have to be applied to the inclined faces of the deformed specimen in order to maintain this deformation (Fig. 3.12). Since $\kappa = \tan \theta$, it follows that on BC the outward unit normal vector $\mathbf{n}$

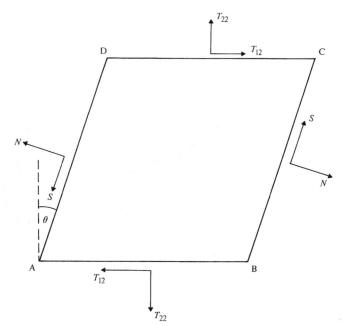

Fig. 3.12

is given by

$$\mathbf{n} = \left(\frac{1}{(1+\kappa^2)^{\frac{1}{2}}}, \frac{-\kappa}{(1+\kappa^2)^{\frac{1}{2}}}, 0 \right) \tag{3.7.9}$$

and the unit tangent vector has components

$$\mathbf{s} = \left(\frac{\kappa}{(1+\kappa^2)^{\frac{1}{2}}}, \frac{1}{(1+\kappa^2)^{\frac{1}{2}}}, 0 \right) \tag{3.7.10}$$

Now

$$N = \mathbf{t(n)} \cdot \mathbf{n} = t_i n_i = T_{ij} n_j n_i$$
$$S = \mathbf{t(n)} \cdot \mathbf{s} = t_i s_i = T_{ij} n_j s_i$$

so that, using (3.7.9) and (3.7.10) we obtain

$$(1+\kappa^2)N = T_{11} + \kappa^2 T_{22} - 2\kappa T_{12}$$
$$(1+\kappa^2)S = \kappa(T_{11} - T_{22}) + (1-\kappa^2)T_{12} \tag{3.7.11}$$

Using (3.7.8), these reduce to

$$N = T_{22} - \kappa S, \qquad S = T_{12}/(1+\kappa^2) \tag{3.7.12}$$

We now see that $S < T_{12}$ and $N < T_{22}$, so that if T_{22} is negative so is N. Also, the Poynting effect is still present when we refer to the actual faces of the sheared cube. For special elastic materials it may be possible for N to be zero for all κ.

3.8 Simple shear of an incompressible material

Using (3.1.14) the stress components in this case take the form

$$T_{11} = -p_0 + 2(1+\kappa^2)W_1 - 2W_2$$
$$T_{22} = -p_0 + 2W_1 - 2(1+\kappa^2)W_2$$
$$T_{33} = -p_0 + 2W_1 - 2W_2 \tag{3.8.1}$$
$$T_{12} = \kappa\mu(\kappa^2), \qquad T_{23} = T_{31} = 0$$

where, as in the previous deformation, p_0 is a constant to be determined by the prescribed boundary conditions, and

$$\mu = 2(W_1 + W_2) \tag{3.8.2}$$

is obtained from (3.7.3) by replacing χ_1 by $2W_1$ and χ_{-1} by

$-2W_2$. Using (3.8.1) in place of (3.7.2), many of the results of the previous section can still be shown to hold; in particular, the remarks concerning the behaviour of normal stresses and shear stresses when the direction of shear is reversed carry over, as do the results (3.7.8), (3.7.11), and (3.7.12). We therefore see that T_{11} and T_{22} cannot be equal, and the Poynting effect is still present.

In contrast with the compressible case, in an incompressible material it is possible to make any one of the normal stresses vanish by an appropriate choice of p_0, and this can be done for any strain-energy function. As an example, consider the situation when the faces of the block normal to the 3-direction are traction-free. Since $T_{31} = T_{32} = 0$ everywhere for this deformation, this condition is satisfied when $T_{33} = 0$ in which case

$$p_0 = 2(W_1 - W_2) \qquad (3.8.3)$$

giving

$$T_{11} = 2\kappa^2 W_1, \quad T_{22} = -2\kappa^2 W_2$$
$$T_{12} = \kappa\mu(\kappa^2), \quad T_{13} = T_{23} = T_{33} = 0 \qquad (3.8.4)$$

Thus W_1 and W_2 may be interpreted directly in terms of normal stresses.

If

$$W_1 > 0, \quad W_2 > 0 \qquad (3.8.5)$$

which are equivalent to inequalities (3.7.6) when the material is incompressible, then $\mu(\kappa^2) > 0$. Further, from (3.8.4) we see that the normal stress on the shearing planes is always a pressure since $T_{22} < 0$. Also, from (3.7.12), we see that $N < 0$, so that the normal stress on each of the deformed faces $X_1 = \text{constant}$ and $X_2 = \text{constant}$ is a pressure. If these pressures are not applied in addition to the shear forces, we would expect the material to stretch in the 1- and 2-directions, and hence contract in the 3-direction (in view of the incompressibility condition). This is the form of the Poynting effect most commonly observed.

3.9 Non-homogeneous deformations

All the deformations considered so far in this chapter are possible in every isotropic elastic material because the form of W has not been specified. Ericksen (1955) has shown that *only* homogeneous deformations are possible in every homogeneous, compressible, isotropic elastic material. In other words, there is no non-homogeneous deformation which is possible in *every* compressible material although, of course, it may be possible in a *particular* material. This is a useful result to the experimentalist who is interested in determining the form of the strain-energy function for a particular material. Only by choosing his deformation to be homogeneous can he be sure that it can take place in the given material. It is no use designing an experiment which would produce a non-homogeneous deformation, such as simple torsion, since that particular deformation may not be a possible one in the material under consideration.

A different situation exists for incompressible, isotropic, elastic materials. The complete set of exact solutions possible in every material is currently unknown. However, several non-homogeneous deformations have been shown to be exact solutions of the basic equations (3.1.2), (3.1.13), and (3.1.14). For such deformations the stress field varies throughout the body and the analysis, usually in curvilinear coordinates, is more involved than for homogeneous deformations. We therefore restrict our attention in this section to the two non-homogeneous deformations discussed in Section 1.6 which are tractable in Cartesian coordinates, and are relevant to experimental work. We assume that body forces can be neglected, although we can deal with a conservative body force, such as gravity, by replacing p in the analysis by $p + \chi$, where χ is the body-force potential ($\rho\mathbf{b} = -\text{grad }\chi$).

3.10 Simple torsion of a circular cylinder. Theory

Using the coordinate system shown in Fig. 1.4, we first consider a circular cylinder of incompressible elastic material, of length l and radius a, subjected to the deformation

$$x_1 = cX_1 - sX_2, \quad x_2 = sX_1 + cX_2, \quad x_3 = X_3 \qquad (3.10.1)$$

where $s = \sin \tau X_3$, $c = \cos \tau X_3$, and τ is a constant. In Section 1.6 we saw that this deformation satisfies the constraint $J = 1$, and since there is no change in shape of the cylinder during the deformation, its length remains l and its radius a. We consider the possibility of maintaining the deformation (3.10.1) by applying surface

tractions to the end sections while the curved surface $r = a$
$(r^2 = x_1^2 + x_2^2)$ remains free of traction.

In discussing this problem it is convenient to express $\mathbf{B}$ and
$\mathbf{B}^{-1}$ in terms of spatial coordinates. Using (3.10.1), (1.6.6), and
(1.6.7), we find

$$\mathbf{B} = \begin{pmatrix} 1 + \tau^2 x_2^2 & -\tau^2 x_1 x_2 & -\tau x_2 \\ -\tau^2 x_1 x_2 & 1 + \tau^2 x_1^2 & \tau x_1 \\ -\tau x_2 & \tau x_1 & 1 \end{pmatrix} \tag{3.10.2}$$

$$\mathbf{B}^{-1} = \begin{pmatrix} 1 & 0 & \tau x_2 \\ 0 & 1 & -\tau x_1 \\ \tau x_2 & -\tau x_1 & 1 + \tau^2(x_1^2 + x_2^2) \end{pmatrix} \tag{3.10.3}$$

The stress components associated with this deformation are now
obtained by substituting (3.10.2) and (3.10.3) into (3.1.14). They
are

$$\begin{aligned}
T_{11} &= -p + 2(1 + \tau^2 x_2^2) W_1 - 2 W_2 \\
T_{22} &= -p + 2(1 + \tau^2 x_1^2) W_1 - 2 W_2 \\
T_{33} &= -p + 2 W_1 - 2(1 + \tau^2 r^2) W_2 \\
T_{12} &= -2\tau^2 x_1 x_2 W_1, \quad T_{23} = \tau\mu(\tau^2 r^2) x_1, \quad T_{31} = -\tau\mu(\tau^2 r^2) x_2
\end{aligned} \tag{3.10.4}$$

where, in view of (1.6.8), the coefficients W_1 and W_2 are functions of $\tau^2 r^2$, and μ is as defined in (3.8.2) with κ^2 replaced by
$\tau^2 r^2$. We note that in this deformation the stresses vary within the
specimen.

As in the earlier deformations, the pressure p is determined by
the equilibrium equations and the boundary conditions. In the
absence of body force, the equilibrium equations (3.1.2) reduce
to

$$T_{ij,j} = 0 \tag{3.10.5}$$

Since W is independent of x_3, $T_{13,3} = T_{23,3} = 0$. Also, using
(3.10.4),

$$\begin{aligned}
T_{31,1} + T_{32,2} &= -\tau x_2 \frac{\partial \mu}{\partial x_1} + \tau x_1 \frac{\partial \mu}{\partial x_2} \\
&= \left(-\tau \frac{x_1 x_2}{r} + \tau \frac{x_1 x_2}{r} \right) \frac{d\mu}{dr} = 0
\end{aligned}$$

and the third equation of (3.10.5), $i = 3$, reduces to

$$\frac{\partial p}{\partial x_3} = 0 \tag{3.10.6}$$

so that p, and hence all the stress components (3.10.4), are independent of x_3. Substituting $(3.10.4)_{1,4,6}$ into the first equation of (3.10.5), $i = 1$, gives

$$\frac{\partial p}{\partial x_1} = \frac{\partial}{\partial x_1} \{2(W_1 - W_2) + 2\tau^2 x_2^2 W_1\} - 2\tau^2 x_1 \frac{\partial}{\partial x_2}(x_2 W_1) \tag{3.10.7}$$

and using $(3.10.4)_{2,4,5}$ in the second equation of (3.10.5), $i = 2$, gives

$$\frac{\partial p}{\partial x_2} = \frac{\partial}{\partial x_2} \{2(W_1 - W_2) + 2\tau^2 x_1^2 W_1\} - 2\tau^2 x_2 \frac{\partial}{\partial x_1}(x_1 W_1) \tag{3.10.8}$$

Now

$$x_1 = r \cos \theta, \qquad x_2 = r \sin \theta \tag{3.10.9}$$

so that, by the chain rule for partial differentiation,

$$\frac{\partial p}{\partial \theta} = \frac{\partial p}{\partial x_1} \frac{\partial x_1}{\partial \theta} + \frac{\partial p}{\partial x_2} \frac{\partial x_2}{\partial \theta} = -x_2 \frac{\partial p}{\partial x_1} + x_1 \frac{\partial p}{\partial x_2}$$

$$= 2 \frac{\partial}{\partial \theta}(W_1 - W_2) + 2\tau^2 r^2 \frac{\partial W_1}{\partial \theta} \tag{3.10.10}$$

using (3.10.7) and (3.10.8). Since W_1 and W_2 are independent of θ, $\partial p / \partial \theta = 0$. Hence, p depends only on r. Further

$$\frac{dp}{dr} = \frac{\partial p}{\partial x_1} \frac{\partial x_1}{\partial r} + \frac{\partial p}{\partial x_2} \frac{\partial x_2}{\partial r} = \frac{x_1}{r} \frac{\partial p}{\partial x_1} + \frac{x_2}{r} \frac{\partial p}{\partial x_2}$$

using (3.10.9), or

$$r \frac{dp}{dr} = x_1 \frac{\partial p}{\partial x_1} + x_2 \frac{\partial p}{\partial x_2} \tag{3.10.11}$$

Substituting (3.10.7) and (3.10.8) into (3.10.11), we find, after a little manipulation,

$$r \frac{dp}{dr} = 2 \left(x_1 \frac{\partial}{\partial x_1} + x_2 \frac{\partial}{\partial x_2} \right)(W_1 - W_2) - 2\tau^2 r^2 W_1$$

giving

$$\frac{dp}{dr} = 2\frac{d}{dr}(W_1 - W_2) - 2\tau^2 r W_1 \qquad (3.10.12)$$

Once W is known, (3.10.12) determines p to within an arbitrary additive constant which is found from the boundary conditions. For example, if we are interested in a Mooney–Rivlin material

$$W_1 = C_1, \qquad W_2 = C_2 \qquad (3.10.13)$$

so that

$$p = -\tau^2 C_1(r^2 - a^2) + p_a \qquad (3.10.14)$$

where p_a is a constant, the value of p on $r = a$, to be determined later from the traction-free boundary condition on $r = a$. On any cylindrical surface, of radius r, the unit normal is given by

$$\mathbf{n} = \frac{1}{r}(x_1, x_2, 0) = \mathbf{e}_r \qquad (3.10.15)$$

so that, using (1.11.11) and (3.10.4), the components of the stress vector on such a surface are given by

$$t_1 = T_{11}\frac{x_1}{r} + T_{12}\frac{x_2}{r} = \{-p + 2(W_1 - W_2)\}\frac{x_1}{r}$$

$$t_2 = T_{21}\frac{x_1}{r} + T_{22}\frac{x_2}{r} = \{-p + 2(W_1 - W_2)\}\frac{x_2}{r} \qquad (3.10.16)$$

$$t_3 = T_{31}\frac{x_1}{r} + T_{32}\frac{x_2}{r} = 0$$

This stress vector is therefore radial and given by

$$\mathbf{t}(\mathbf{e}_r) = \{-p + 2(W_1 - W_2)\}\mathbf{e}_r \qquad (3.10.17)$$

For the curved surface of the cylinder to be traction-free we require

$$\mathbf{t}(\mathbf{e}_r) = \mathbf{0} \qquad (3.10.18)$$

on $r = a$, which is satisfied if

$$p = 2(W_1 - W_2) \qquad (3.10.19)$$

on $r = a$. Combining (3.10.12) and (3.10.19) we see that the value of p which satisfies both the equilibrium equations and traction-free boundary condition is

$$p = 2\tau^2 \int_r^a sW_1 \, ds + 2(W_1 - W_2) \qquad (3.10.20)$$

where in the integrand W_1 is written as a function of s in place of r. The expression for the stress vector $\mathbf{t}(\mathbf{e}_r)$ now simplifies to

$$\mathbf{t}(\mathbf{e}_r) = \left(-2\tau^2 \int_r^a sW_1 \, ds\right)\mathbf{e}_r \qquad (3.10.21)$$

In the particular case of a Mooney–Rivlin material, using (3.10.13), these become

$$\begin{aligned} p &= -\tau^2 C_1(r^2 - a^2) + 2(C_1 - C_2) \\ \mathbf{t}(\mathbf{e}_r) &= -\tau^2 C_1(a^2 - r^2)\mathbf{e}_r \end{aligned} \qquad (3.10.22)$$

We now consider the boundary conditions which must be applied to the end sections in order to maintain this deformation. Since the stresses are independent of x_3, there is no need to consider each end section separately. Using (3.10.4) and (3.10.20),

$$\mathbf{t}(\mathbf{e}_3) = (T_{31}, T_{32}, T_{33}) = \left(-\mu\tau x_2, \mu\tau x_1, -2\tau^2 \int_r^a sW_1 \, ds - 2\tau^2 r^2 W_2\right) \qquad (3.10.23)$$

Since

$$\int_S -\tau\mu(\tau^2 r^2)x_2 \, dx_1 \, dx_2 = -\tau \int_0^a \mu(\tau^2 r^2)r^2 \, dr \int_0^{2\pi} \sin\theta \, d\theta = 0$$

and

$$\int_S \tau\mu x_1 \, dx_1 \, dx_2 = 0$$

where S denotes the region $x_1^2 + x_2^2 \leq a^2$, we see that the resultant force on the end section has zero components in the plane of the section. Further, the normal component N of the resultant force is given by

$$\begin{aligned} N &= \int_S T_{33} \, dx_1 \, dx_2 = \int_0^{2\pi} \int_0^a \left\{-2\tau^2 \int_r^a sW_1 \, ds - 2\tau^2 r^2 W_2\right\} r \, dr \, d\theta \\ &= -2\pi\tau^2 \int_0^a \left\{2r \int_r^a sW_1 \, ds + 2r^3 W_2\right\} dr \end{aligned}$$

But

$$\int_0^a r\left(\int_r^a sW_1\,ds\right)dr = \left[\tfrac{1}{2}r^2\int_r^a sW_1\,ds\right]_0^a + \tfrac{1}{2}\int_0^a r^3W_1\,dr$$

$$= \left[\frac{r^2}{4\tau^2}\,\mathbf{t}(\mathbf{e}_r)\cdot\mathbf{e}_r\right]_0^a + \tfrac{1}{2}\int_0^a r^3W_1\,dr$$

$$= \tfrac{1}{2}\int_0^a r^3W_1\,dr$$

since $r^2\mathbf{t}(\mathbf{e}_r)\cdot\mathbf{e}_r \to 0$ as $r \to 0$, and $\mathbf{t}(\mathbf{e}_r)=\mathbf{0}$ when $r=a$. Thus

$$N = -2\pi\tau^2\int_0^a r^3(W_1+2W_2)\,dr \qquad (3.10.24)$$

In addition to this normal force, the components of stress produce a couple about the axis of the cylinder of magnitude

$$M = \int_S (x_1 T_{23} - x_2 T_{13})\,dS$$

$$= \int_0^{2\pi}\int_0^a \mu\tau r^3\,dr\,d\theta$$

$$= 2\pi\tau\int_0^a r^3\mu(\tau^2 r^2)\,dr \qquad (3.10.25)$$

Thus M is determined entirely by the form of the generalised shear modulus μ.

Example 3.10.1

Write down expressions for the moments about the 1- and 2-directions and show that they vanish.

For a Mooney–Rivlin material the expressions (3.10.24) and (3.10.25) become

$$N = -\tfrac{1}{2}(C_1+2C_2)\pi\tau^2 a^4, \qquad M = (C_1+C_2)\pi\tau a^4 \qquad (3.10.26)$$

The ratio M/τ is called the *torsional rigidity*. From (3.10.25) we see that this ratio is an even function of τ, and from (3.10.24) we see that N is also an even function of τ.

The limits of M/τ and N/τ^2, as $\tau \to 0$, are both constant, and this is of significance in the discussion of the experimental work.

In fact, from (3.10.24) and (3.10.25), we see that

$$\lim_{\tau \to 0} \frac{N}{\tau^2} = -\tfrac{1}{2}\pi a^4 (W_1^0 + 2W_2^0)$$

$$\lim_{\tau \to 0} \frac{M}{\tau} = \tfrac{1}{2}\pi \mu(0) a^4$$

(3.10.27)

where W_1^0, W_2^0 denote W_1, W_2 evaluated with $\tau = 0$.

When the inequalities (3.8.5) are satisfied we have $N < 0$, so that the resultant force on the end sections required to maintain the simple torsion is always a pressure. If this force is not applied, the specimen will tend to elongate. As mentioned in Section 3.7 such a phenomenon was observed by Poynting (1909). In addition he found that the elongation was proportional to the square of the twist τ. While the above analysis shows that to prevent such an elongation, a pressure has to be applied, to deal with Poynting's actual experiment we must consider a torsional deformation together with an axial extension and radial contraction. This is discussed in Sections 3.12 and 3.13.

3.11 Simple torsion of a circular cylinder. Experiment

The mathematical prediction that, in order to produce simple torsion, a normal force must be applied to the end sections of a cylinder in addition to a couple, was confirmed experimentally by Rivlin (1947).

In a preliminary experiment, a rubber cylinder of pure gum compound, with brass plates A and B bonded on to each plane end, was subjected to a torsional couple by means of a lever attached to one of the ends while the other end was held fixed in a vice. Unless a thrust was applied parallel to the axis of the cylinder, the cylinder elongated, its axial cross-section before and after torsion being shown roughly in Fig. 3.13.

For a material with the Mooney form of strain-energy function, it follows from (2.9.3) and (3.10.23) that the normal traction at different points on the end section is given by

$$T_{33} = -2\tau^2 a^2 C_2 - \tau^2 (a^2 - r^2)(C_1 - 2C_2)$$

(3.11.1)

This stress is therefore distributed parabolically over the surface,

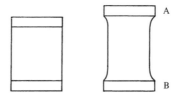

Fig. 3.13   From Rivlin, 1947

and, in the particular case when $C_2 = 0$, vanishes at the circumference. By measuring T_{33} over the plane ends of the cylinder, Rivlin obtained an estimate of the ratio C_2/C_1.

The apparatus used consisted of a rubber cylinder with plane ends, to which brass discs were welded. One disc was then rotated relative to the other while their distance apart remained fixed, thus preventing any elongation of the cylinder. One of the discs contained five circular holes, varying in distance from the centre of the disc. When the other disc was rotated, the rubber bulged slightly in the holes. The height of each bulge was assumed to be proportional to the thrust which would have to be exerted to prevent the bulging.

If d denotes the height of the bulge in any one of the holes measured at the centre for various angles θ of rotation of the disc B then, under the above assumption, using (3.11.1),

$$d = K\theta^2\{2a^2C_2 + (a^2 - r^2)(C_1 - 2C_2)\} \qquad (3.11.2)$$

where K is some constant of proportionality. If (3.11.2) is applicable, a graph in which d is plotted against θ^2 should be a straight line passing through the origin. This was found to be the case for each hole within the accuracy of the experiment. Now for the hole at distance r from the centre of the cylinder, the slope m of the graph of d against θ^2 is $K\{2a^2C_2 + (a^2 - r^2)(C_1 - 2C_2)\}$, and if m is plotted against $(a^2 - r^2)$, a further straight line should result with intercept $2a^2C_2K$ on the m-axis. Again this was verified from the experimental data, and from the value of the intercept and the slope $K(C_1 - 2C_2)$, Rivlin obtained the value $1/7$ for the ratio C_2/C_1, in good agreement with the value of W_2/W_1 obtained in the experiments described in Section 3.6(ii).

In a subsequent more exact study, Rivlin and Saunders measured the resultant couple M and force N on the plane ends. For the special case of a Mooney–Rivlin strain energy, we see from

(3.10.26) that M is proportional to the amount of twist τ, and the total thrust $-N$ to the square of the twist. If the measurements of M and $-N$ are made for various values of τ, then the slope of the graph of M against τ gives a value of $C_1 + C_2$, and that of the graph of $-N$ against τ^2 gives a value of $C_2 + \frac{1}{2}C_1$. From these values C_2/C_1 can be calculated. If either of these curves departs from a straight line this will indicate a departure of the strain-energy function from Mooney form.

The experimental arrangement used by Rivlin and Saunders is shown in Fig. 3.14. The rubber cylinder was attached, by the plane ends bonded to it, to the circular plate P and the base plate B. A known torque was applied to the rubber cylinder by strings supporting equal weights W; the torque was varied by altering these weights. As the torque increased from zero, the cylinder elongated, and this elongation was observed by noting the rise or fall of two points on the edge of the plate P at opposite ends of a diameter. The plate was kept horizontal by a suitable placing of the mass m. At each value of the applied torque the magnitude of the counterpoise L which would result in no extension of the cylinder was found by measuring the extension and compression for values of L slightly greater and slightly less than this value

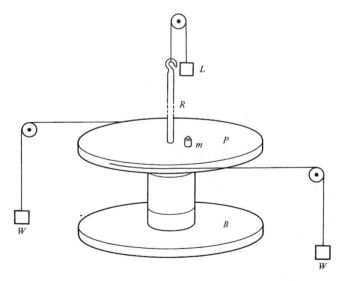

Fig. 3.14   From Rivlin and Saunders, 1951

and interpolating. The difference between this value of L and that obtained for zero torque gives the value of the thrust $-N$ corresponding to the particular value of the applied torque employed. The amount of torsion produced was measured by reading a scale on the edge of the plate P against a fiduciary mark.

The results obtained are shown in Figs. 3.15 and 3.16. In each case curve I corresponds to the torque being steadily increased, and curve II to a steady decrease to zero. Apart from a slight curvature towards the τ-axis, the graph of M against τ is very nearly linear. The relation between $-N$ and τ^2 is also seen to be linear within the accuracy of the experiment. From (3.10.24) we see that whatever the form of the strain-energy function, the curve of $-N$ against τ^2 should pass through the origin whereas, from the experimental results, both curves have a finite intercept. The equations used in deriving (3.10.24) are for an elastic material which is isotropic in its undeformed state. If the material were assumed to be anisotropic in its undeformed state, equation

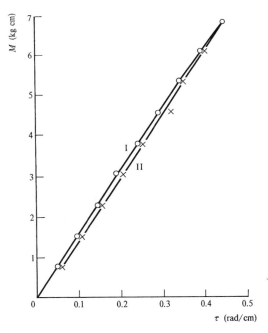

Fig. 3.15   From Rivlin and Saunders, 1951

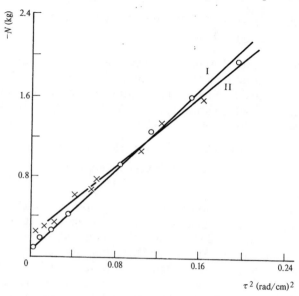

Fig. 3.16   From Rivlin and Saunders, 1951

(3.10.24) would be modified. Any hysteresis or permanent set associated with the higher values of τ would result in some anisotropy, and this may explain the non-zero intercept obtained.

The approximate linearity of the two curves suggests that W_1 and W_2 are very nearly constant over the range of values of I_1 and I_2 covered by the experiment, so that the Mooney form (2.9.3) of the strain-energy function is approximately valid, particularly for small values of τ. As can be seen from (1.6.8) the values of I_1 and I_2 vary with the radius from the value 3 on the axis to a maximum on the curved surface. In this experiment the values on the curved surface ranged from 3 to 3.32 so that the values of $I_1 - 3$ and $I_2 - 3$ are less than $\frac{1}{3}$. By considering the slopes of the curves, as mentioned earlier, an approximate value of C_2/C_1 can be found. Using the curves labelled I, the calculation gives the value $\frac{1}{3}$. It should be noted that the specimen used here differs in composition from that used in the experiments described earlier in this section. In fact, when Rivlin's earlier experiment was repeated with this specimen, a similar value was obtained for C_2/C_1.

3.12   Extension and torsion of a circular cylinder. Theory

Finally, we consider a hollow circular cylinder of incompressible material with initial length l and initial external and internal radii a and b, respectively. Using the coordinate system shown in Fig. 1.4 we take the deformation to be (1.6.11) subject to the incompressibility condition (1.6.14), that is,

$$
\begin{aligned}
x_1 &= \alpha^{-\frac{1}{2}}\{X_1 \cos(\alpha\tau X_3) - X_2 \sin(\alpha\tau X_3)\} \\
x_2 &= \alpha^{-\frac{1}{2}}\{X_1 \sin(\alpha\tau X_3) + X_2 \cos(\alpha\tau X_3)\} \qquad (3.12.1) \\
x_3 &= \alpha X_3
\end{aligned}
$$

This isochoric deformation represents a simple extension together with a simple torsion. Since there is no volume change, there is a change of radii, that is a deflation, associated with the extension. From (3.12.1), with $\tau = 0$, we see that after simple extension the internal and external radii of the tube are $\alpha^{-\frac{1}{2}}b$ and $\alpha^{-\frac{1}{2}}a$ respectively, and the length is αl. As we have seen in the previous section, these are unchanged by torsion. We suppose that the outer curved surface is traction-free, and we determine the tractions which have to be applied to the remaining surfaces to maintain this deformation.

The tensors $\mathbf{B}$ and $\mathbf{B}^{-1}$, in terms of spatial coordinates, take the forms

$$
\mathbf{B} = \begin{pmatrix}
\alpha^{-1} + (\alpha\tau x_2)^2 & -\alpha^2\tau^2 x_1 x_2 & -\alpha^2\tau x_2 \\
-\alpha^2\tau^2 x_1 x_2 & \alpha^{-1} + (\alpha\tau x_1)^2 & \alpha^2\tau x_1 \\
-\alpha^2\tau x_2 & \alpha^2\tau x_1 & \alpha^2
\end{pmatrix} \qquad (3.12.2)
$$

$$
\mathbf{B}^{-1} = \begin{pmatrix}
\alpha & 0 & \alpha\tau x_2 \\
0 & \alpha & -\alpha\tau x_1 \\
\alpha\tau x_2 & -\alpha\tau x_1 & \alpha^{-2} + \alpha\tau^2(x_1^2 + x_2^2)
\end{pmatrix} \qquad (3.12.3)
$$

and the stress components are given by

$$
\begin{aligned}
T_{11} &= -p + 2\{\alpha^{-1} + (\alpha\tau x_2)^2\}W_1 - 2\alpha W_2 \\
T_{22} &= -p + 2\{\alpha^{-1} + (\alpha\tau x_1)^2\}W_1 - 2\alpha W_2 \\
T_{33} &= -p + 2\alpha^2 W_1 - 2(\alpha^{-2} + \tau^2 R^2)W_2 \\
T_{12} &= -2\alpha^2\tau^2 x_1 x_2 W_1 \qquad\qquad\qquad (3.12.4) \\
T_{23} &= \tau\bar{\mu}(\alpha, \tau^2 R^2)x_1 \\
T_{31} &= -\tau\bar{\mu}(\alpha, \tau^2 R^2)x_2
\end{aligned}
$$

where

$$\bar{\mu} = 2\alpha(\alpha W_1 + W_2), \qquad R^2 = \alpha r^2 \tag{3.12.5}$$

We see from (1.6.17) that W_1 and W_2 are functions of α and $\tau^2 R^2$.

As usual, we have to determine p so that the equilibrium equations (3.10.5) are satisfied. By working similar to that given in Section 3.10, but with μ replaced by $\bar{\mu}$, it follows from $(3.12.4)_{5,6}$ that the third equilibrium equation reduces to (3.10.6) so that p, and all the stress components (3.12.4), are independent of x_3. Using (3.12.4) the remaining equilibrium equations reduce to

$$\frac{\partial p}{\partial x_1} = \frac{\partial}{\partial x_1}\{2(\alpha^{-1}W_1 - \alpha W_2) + 2(\alpha\tau)^2 x_2^2 W_1\} - 2(\alpha\tau)^2 x_1 \frac{\partial}{\partial x_2}(x_2 W_1)$$

$$\frac{\partial p}{\partial x_2} = \frac{\partial}{\partial x_2}\{2(\alpha^{-1}W_1 - \alpha W_2) + 2(\alpha\tau)^2 x_1^2 W_1\} - 2(\alpha\tau)^2 x_2 \frac{\partial}{\partial x_1}(x_1 W_1)$$

$$\tag{3.12.6}$$

Example 3.12.1

Using (3.12.6) and proceeding as in Section 3.10, show that p depends upon r only, and that

$$\frac{dp}{dr} = 2\frac{d}{dr}(\alpha^{-1}W_1 - \alpha W_2) - 2(\alpha\tau)^2 r W_1 \tag{3.12.7}$$

Example 3.12.2

Show that, on any surface $r = $ constant, the stress vector is given by

$$\mathbf{t}(\mathbf{e}_r) = (-p + 2\alpha^{-1}W_1 - 2\alpha W_2)\mathbf{e}_r \tag{3.12.8}$$

Since after deformation the internal and external radii of the tube are $\alpha^{-\frac{1}{2}}b$ and $\alpha^{-\frac{1}{2}}a$ respectively, when discussing boundary conditions in the current configuration we are interested in the boundaries $r = \alpha^{-\frac{1}{2}}b$ and $r = \alpha^{-\frac{1}{2}}a$. If the outer curved surface is to be traction-free we require

$$p = 2\alpha^{-1}W_1 - 2\alpha W_2 \tag{3.12.9}$$

on $r = a\alpha^{-\frac{1}{2}}$.

Example 3.12.3

Use (3.12.7), (3.12.8), and (3.12.9) to show that

$$\mathbf{t}(\mathbf{e}_r) = \left(-2\alpha^2\tau^2 \int_r^{a\alpha^{-\frac{1}{2}}} sW_1 \, ds\right)\mathbf{e}_r \qquad (3.12.10)$$

where W_1 is written as a function of s in the place of r.

In order to prevent further change of the inner radius during the torsion, we consider the application of a uniform pressure P on the inner surface. So on that boundary we have

$$\mathbf{t}(-\mathbf{e}_r) = P\mathbf{e}_r \qquad (3.12.11)$$

Using $(3.12.5)_2$ and (3.12.10) it follows that

$$P = 2\alpha\tau^2 \int_b^a RW_1 \, dR \qquad (3.12.12)$$

where W_1 is now written as a function of R. For a Mooney–Rivlin material this gives

$$P = \alpha\tau^2 C_1(a^2 - b^2) \qquad (3.12.13)$$

We see from (3.12.12) that an internal pressure P is necessary to maintain this deformation whenever $W_1 \neq 0$ over the whole range $a > R > b$ and $\tau \neq 0$. If this pressure is not applied, the inner boundary is traction free and the inner radius changes during torsion. The deformation describing this more complicated situation is not discussed in this text.

Example 3.12.4

Using working similar to that in Section 3.10, show that the resultant force on the end section has zero components in the plane of the section, and a normal component N given by

$$N = 4\pi(\alpha - \alpha^{-2}) \int_b^a R(W_1 + \alpha^{-1}W_2) \, dR$$
$$- 2\pi\tau^2 \int_b^a R^3(W_1 + 2\alpha^{-1}W_2) \, dR \qquad (3.12.14)$$

Show further that the stress distribution (3.12.4) also gives rise to a couple M about the axis of the cylinder given by

$$M = 4\pi\tau \int_b^a R^3(W_1 + \alpha^{-1}W_2)\,\mathrm{d}R \qquad (3.12.15)$$

It follows from (1.6.17) that, as $\tau \to 0$, I_1 and I_2 tend to values independent of R. Hence, W_1 and W_2 have constant limiting values as $\tau \to 0$, and from (3.12.12), (3.12.14), and (3.12.15) we find

$$\lim_{\tau \to 0}(M/\tau) = \pi(W_1^0 + \alpha^{-1}W_2^0)(a^4 - b^4)$$

$$\lim_{\tau \to 0} N = 2\pi(\alpha - \alpha^{-2})(W_1^0 + \alpha^{-1}W_2^0)(a^2 - b^2) \qquad (3.12.16)$$

$$\lim_{\tau \to 0}(P/\tau^2) = \alpha W_1^0(a^2 - b^2)$$

where W_1^0 and W_2^0 denote W_1 and W_2 evaluated with $\tau = 0$ and

$$I_1 = \alpha^2 + 2\alpha^{-1}, \qquad I_2 = 2\alpha + \alpha^{-2} \qquad (3.12.17)$$

These results are useful when discussing experiments involving sufficiently small values of the twist τ.

When the specimen is a solid cylinder, the force and couple required to produce the deformation may be found by putting $b = 0$ in (3.12.14) and (3.12.15). Also, the right-hand side of (3.12.12) with $b = 0$ gives the pressure on the axis of the cylinder resulting from the deformation. From (3.12.16) we see that in this case

$$\lim_{\tau \to 0}(M/\tau) = \pi(W_1^0 + \alpha^{-1}W_2^0)a^4$$

$$\lim_{\tau \to 0} N = 2\pi(\alpha - \alpha^{-2})(W_1^0 + \alpha^{-1}W_2^0)a^2 \qquad (3.12.18)$$

From (3.12.15) and (3.12.18) we see that the torsional rigidity M/τ is affected by the extension however small the twist. For a Mooney–Rivlin material ($W_1^0 = C_1$, $W_2^0 = C_2$), the torsional rigidity decreases as α increases and increases when α decreases, so that extension has the effect of softening an incompressible material in torsion while contraction stiffens it.

From (3.12.18) we find

$$\frac{\lim\limits_{\tau \to 0} Na^2}{\lim\limits_{\tau \to 0} (M/\tau)} = 2(\alpha - \alpha^{-2}) \qquad (3.12.19)$$

Since this result, first obtained by Rivlin (1949) is independent of the strain-energy function, it is a universal relation valid for all isotropic, incompressible, elastic materials. Since $d(\alpha - \alpha^{-2})/d\alpha > 0$, the right-hand side of (3.12.19) is an increasing function of α, so that the greater the stretch, the greater must be the stretching forces in proportion to the torsional rigidity.

As mentioned in Section 3.10, Poynting observed that under a torsional couple a solid cylinder experiences an elongation proportional to the square of the twist. This corresponds to $N = 0$, in which case from (3.12.14) we have

$$2(\alpha - \alpha^{-2}) \int_0^a R(W_1 + \alpha^{-1}W_2)\,dR = \tau^2 \int_0^a R^3(W_1 + 2\alpha^{-1}W_2)\,dR$$

$$(3.12.20)$$

This is an equation relating the stretch α to the twist τ. Although in general this equation may not have a solution for α in terms of τ, it follows from (3.12.20) that when a solution does exist, and the inequalities (3.8.5) hold, then $\alpha^3 - 1 > 0$. Hence, $\alpha > 1$ which means that the cylinder always elongates when twisted. Also from (3.12.20), we see that, as $\tau \to 0$, $\alpha \to 1$ and

$$\begin{aligned}
\lim_{\tau \to 0} \frac{\alpha - 1}{\tau^2} &= \lim_{\tau \to 0} \frac{\alpha^3 - 1}{(\alpha^2 + \alpha + 1)\tau^2} \\
&= \frac{1}{3} \lim_{\tau \to 0} \frac{\alpha^3 - 1}{\tau^2} \\
&= \frac{1}{6} \lim_{\tau \to 0} \left\{ \frac{\displaystyle\int_0^a R^3(W_1 + 2\alpha^{-1}W_2)\,dR}{\displaystyle\int_0^a R(W_1 + \alpha^{-1}W_2)\,dR} \right\} \\
&= \frac{a^2}{12} \left\{ \frac{W_1 + 2W_2}{W_1 + W_2} \right\}_{\tau = 0, \alpha = 1} \qquad (3.12.21)
\end{aligned}$$

This expression, which is constant, is sometimes called the *Poynting modulus*. When the inequalities (3.8.5) hold, the Poynting modulus is positive, and since $W_1 + 2W_2 > W_1 + W_2$, this modulus exceeds $a^2/12$.

3.13 Extension and torsion of a circular cylinder. Experiment

(i) Solid cylinder

In order to test relations (3.12.18) and to check the form of the strain-energy function (3.6.8) experimentally, Rivlin and Saunders considered the case of a simple extension followed by a small amount of torsion. At each value of the extension ratio α, the dependence of the torque M on τ was measured. The longitudinal force $[N]_{\tau=0}$ necessary to produce the simple extension under zero torsion was also measured for each value of α. The manner in which M depended on τ for various values of α is shown in Fig. 3.17. It can be seen that for lower values of α the curves depart from linearity, the departure becoming less with increase in α. The dotted lines are the straight lines $M = \tau[M/\tau]_{\tau=0}$. These were obtained using (3.12.18) since knowing $[N]_{\tau=0}$, the total load required to produce various extensions at zero torsion, we can calculate $W_1^0 + \alpha^{-1}W_2^0$ which in turn, using (3.12.18)₁, may be used to calculate $[M/\tau]_{\tau=0}$. The departure of the graph of M

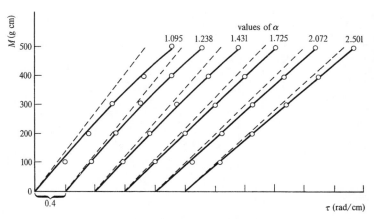

Fig. 3.17 From Rivlin and Saunders, 1951

against τ from a straight line is certainly not accounted for by a Mooney–Rivlin form of strain-energy function, but it can be explained by the fall of $W_1 + \alpha^{-1} W_2$ with increasing deformation. Using these results, as well as some for lower values of τ, Rivlin and Saunders found that the values of $2(\alpha - \alpha^{-2})[M/\tau]_{\tau=0}/a^2[N]_{\tau=0}$ all lie in the range 0.98 to 1.00, in good agreement with (3.12.19).

(ii) Hollow cylinder

From equations (3.12.16) we see that for a given value of α, and sufficiently small values of τ, there should be a linear relation between: (a) the torsional couple M and the amount of torsion τ; and (b) the inflating pressure P, which must be applied to prevent any decrease in the diameter of the tube on twisting, and τ^2.

The experimental arrangement used by Gent and Rivlin (1952) to verify this dependence is shown schematically in Fig. 3.18. The rubber tube T was mounted in a framework F of steel rods by means of its end-plates E_1 and E_2, one end-plate E_1 being fixed.

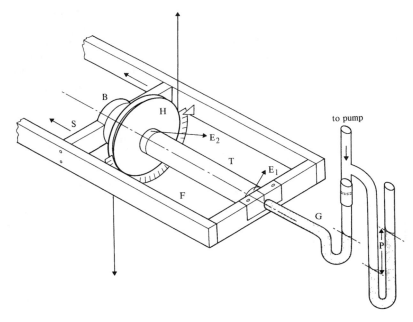

Fig. 3.18   From Gent and Rivlin, 1952. © The Institute of Physics

The other end-plate E_2 was held through a tension ball-race B so that it was free to rotate about the axis of the tube. The ball-race B was in turn held in a slide S by means of which the tube could be extended and clamped in any desired state of extension. The torsion was produced by hanging equal weights to strings passing over the pulley H attached to the end-plate E_2. A glass tube G was connected to the end-plate E_1, communicating with the interior of the rubber tube through a hole in E_1. The rubber tube with the attached glass tube was filled with water before mounting in the framework. After mounting, a water manometer and bicycle pump were connected to the free limb of the tube G.

In performing the experiments, the rubber tube was extended and the extension ratio α was determined by measuring the distances in the undeformed and deformed states between two circumferential lines on it sufficiently far from the ends for end effects to be avoided. The tube was then subjected to various amounts of torsion by adjusting the torsional couple on the pulley. The volume of the tube was maintained constant by adjusting the air pressure above the free surface of the liquid in the tube G by means of the bicycle pump. For each value of the torsional couple the corresponding pressure was measured by means of the water manometer and the relative rotation of the end-plates by means of the scale attached to E_2. From this value of rotation and the length of the tube, the amount of torsion was calculated. The experiments were repeated for three tubes with different diameters.

The values of M, P, and τ obtained experimentally produced accurate linear relations for each value of α employed over the range of values of τ (0 to 0.15 rad/cm approximately) covered by the experiments.

The slopes of the straight lines were used to verify the form of the strain-energy function proposed by Rivlin and Saunders. In particular, from the slope of the (P, τ^2) curve, together with the radii a and b, $(3.12.16)_3$ can be used to calculate W_1^0 corresponding to each value of α. It was found that over the limited range of extension ratios examined these values were independent of α, thus indicating that W_1^0 is independent of I_1 and I_2.

Equation $(3.12.16)_2$ can be used to determine the general mode of variation of W_2^0 with I_2. This dependence was seen to be similar to that found by Rivlin and Saunders using their results with the solid cylinder.

Examples 3

1. Calculate the stress components in a compressible elastic material subjected to the deformation

$$x_1 = \alpha X_1 + \kappa X_2, \quad x_2 = X_2, \quad x_3 = X_3,$$

where α and κ are positive constants, and verify that in the absence of body force the equilibrium equations are satisfied so that this is a possible static deformation. Show that $(T_{11} - T_{22})/T_{12}$ is independent of the response functions. For what values of κ can values of α be found to make $T_{11} = T_{22}$?

2. Interpret geometrically the deformation

$$x_1 = \beta X_1 + \kappa X_2, \quad x_2 = \alpha X_2, \quad x_3 = \beta X_3,$$

where α, β, and κ are positive constants. Show that the stress components in a compressible elastic material subjected to this deformation satisfy the universal relation

$$T_{11} - T_{22} = (\beta^2 + \kappa^2 - \alpha^2)T_{12}/\alpha\kappa$$

3. Calculate the stress components in a compressible elastic body subjected to the deformation

$$x_1 = X_1 + \kappa X_2, \quad x_2 = X_2 + \kappa X_3, \quad x_3 = X_3$$

where κ is a non-zero constant, and deduce that they satisfy the universal relations

$$T_{22} - T_{11} = T_{13} = (T_{12} - T_{23})/\kappa$$

$$T_{22} - T_{33} = T_{13} + \kappa T_{23}$$

Show also that the deformation is isochoric.

4. When undeformed, a rectangular sheet of foam rubber occupies the region $-A \leqslant X_1 \leqslant A$, $-B \leqslant X_2 \leqslant B$, $-H \leqslant X_3 \leqslant H$. The sheet is stretched in the 1-direction by applying a uniform tension T per unit area of the deformed configuration to the edges normal to the 1-direction. The other edges are constrained to remain in the planes $X_2 = \pm B$, and the major faces of the sheet (orthogonal to the 3-direction) are traction-free. Assuming that the deformation is given by

$$x_1 = \lambda X_1, \quad x_2 = X_2, \quad x_3 = \Lambda X_3$$

where λ and Λ are positive constants, and that the foam rubber is characterised by the strain-energy function

$$W = \frac{\mu}{2}\left\{\frac{I_2}{I_3} + 2I_3^{\frac{1}{2}} - 5\right\}$$

(cf. (2.9.10)), in which μ is a positive constant, verify that, in the absence of body force, the stretched sheet is in equilibrium. Show that $\lambda = \Lambda^{-3}$, and deduce that if $T < \mu$ the thickness of the deformed sheet is $2H\{1 - (T/\mu)\}^{\frac{1}{3}}$.

5. A material, sometimes referred to as a compressible neo-Hookean rubber, is characterised by a strain-energy function

$$W = \frac{\mu}{2}(I_1 - 3I_3^{\frac{1}{3}}) + \frac{k}{m}\left(I_3^{\frac{1}{2}} - \frac{m}{m-1} + \frac{I_3^{\frac{1}{2}(1-m)}}{m-1}\right)$$

where μ (>0), k (>0), and m are constants. Derive the equations for the stress components in such a material.

A spherical ball made of this material has radius A before deformation, and is subjected to a uniform compression which reduces the radius to a. The deformation is given by

$$x_1 = \lambda X_1, \quad x_2 = \lambda X_2, \quad x_3 = \lambda X_3$$

where λ is a constant, and the material and spatial coordinates X_i, x_i are referred to common rectangular Cartesian axes through the centre of the ball. If the ball is held in equilibrium, in the absence of body force, obtain a relation between a and the pressure p acting on the surface of the ball.

6. Undeformed Mooney–Rivlin material occupies the region $X_1^2 + X_2^2 \le A^2$, $0 \le X_3 \le L$, where A and L are constants. The material is subjected to the deformation

$$x_1 = X_1, \quad x_2 = X_2, \quad x_3 = X_3 + \kappa(X_1^2 + X_2^2)$$

where κ is a positive constant. Find the associated stress field and show that the equilibrium equations are satisfied, in the absence of body force, provided

$$p = p_0 + 8\kappa(C_1 + C_2)x_3 - 12\kappa^2 C_2(x_1^2 + x_2^2),$$

where p_0 is a constant.

If, in the deformed configuration, the traction on the surface $X_1^2 + X_2^2 = A^2$ at the intersection with $X_3 = 0$ is entirely in the

3-direction, find p_0. Show that elsewhere on the surface $X_1^2 + X_2^2 = A^2$, the normal stress must be compressive with magnitude increasing linearly with X_3. Show that the tangential traction is constant and find the resultant force on this surface.

7. An incompressible rubber cylinder which, in its undeformed state, has radius A and length L, is rotated about its axis with constant angular speed Ω. A constant extension (or contraction) is also maintained in the axial direction with a corresponding lateral contraction (or extension). Relative to a set of axes in which the 3-direction coincides with the axis of rotation, the motion is given by

$$x_1 = \alpha^{-\frac{1}{2}}(X_1 \cos \Omega t - X_2 \sin \Omega t),$$

$$x_2 = \alpha^{-\frac{1}{2}}(X_1 \sin \Omega t + X_2 \cos \Omega t), \quad x_3 = \alpha X_3$$

(cf. Example 1 at the end of Chapter 1) where α is a positive constant. Assuming that the rubber is a Mooney–Rivlin material, show that, in the absence of body force, the equations of motion reduce to

$$\frac{\partial p}{\partial x_1} = \rho \Omega^2 x_1, \quad \frac{\partial p}{\partial x_2} = \rho \Omega^2 x_2, \quad \frac{\partial p}{\partial x_3} = 0$$

and hence determine the pressure p.

If the curved surface is traction-free, determine the distribution of stress within the material, and show that if the resultant force on each end section is zero, α must satisfy the equation

$$\alpha^4 + k\alpha^3 + (n^2 - 1)\alpha - k = 0$$

where

$$k = C_2/C_1, \quad n^2 = \rho \Omega^2 A^2 / 8 C_1$$

Show that only one positive root exists, and that this corresponds to an axial contraction.

8. Find $\mathbf{B}$ and $\mathbf{B}^{-1}$ in terms of x_1, x_2, and x_3 for the deformation

$$x_1 = \left(\frac{2X_1}{A} + B\right)^{\frac{1}{2}} \cos AX_2, \quad x_2 = \left(\frac{2X_1}{A} + B\right)^{\frac{1}{2}} \sin AX_2, \quad x_3 = X_3$$

where A and B are constants. Hence, show that the strain invariants are given by

$$I_1 = I_2 = 1 + A^2 r^2 + 1/A^2 r^2, \quad I_3 = 1,$$

where $r^2 = x_1^2 + x_2^2$.

Show that, in an incompressible, isotropic, elastic body, the stress components associated with the deformation are given by

$$T_{11} + T_{22} = -2p - 2W_1(A^2r^2 + 1/A^2r^2)$$

$$T_{11} - T_{22} + 2iT_{12} = -re^{2i\theta}\, \mathrm{d}W/\mathrm{d}r$$

$$T_{33} = -p + 2(W_1 - W_2), \qquad T_{31} = T_{32} = 0$$

where $x_1 + ix_2 = re^{i\theta}$.

In the absence of body force, show that the equilibrium equations are equivalent to

$$\left(\frac{\partial}{\partial x_1} + i\frac{\partial}{\partial x_2}\right)(T_{11} + T_{22}) + \left(\frac{\partial}{\partial x_1} - i\frac{\partial}{\partial x_2}\right)(T_{11} - T_{22} + 2iT_{12}) = 0$$

and verify that

$$\frac{\partial}{\partial r} + \frac{i}{r}\frac{\partial}{\partial \theta} = e^{-i\theta}\left(\frac{\partial}{\partial x_1} + i\frac{\partial}{\partial x_2}\right)$$

Deduce that

$$p = -W - 2A^2r^2W_2 + 2W_1/A^2r^2 - K$$

where K is a constant.

Using the relations (cf. Example 8 at the end of Chapter 1)

$$T_{rr} + T_{\theta\theta} = T_{11} + T_{22}, \quad T_{rr} - T_{\theta\theta} + 2iT_{r\theta} = e^{-2i\theta}(T_{11} - T_{22} + 2iT_{12})$$

show that

$$T_{rr} = W + K, \quad T_{\theta\theta} = \frac{\mathrm{d}}{\mathrm{d}r}(rW) + K, \quad T_{r\theta} = 0$$

$$T_{33} = \frac{\mathrm{d}}{\mathrm{d}r}(rW) + 2(W_1 - W_2) + K - 2A^2r^2W_1 + \frac{2W_2}{A^2r^2}$$

A certain rectangular region in the 1,2-plane of $\mathscr{C}_0$ deforms into an annular sector $r_1 \leqslant r \leqslant r_2$, $-\theta_0 \leqslant \theta \leqslant \theta_0$ (Example 5 of Chapter 1). Show that, if K is suitably chosen, the surface $r = r_1$ is free of traction. Deduce that $r = r_2$ is also free of traction if $r_2 = 1/A^2r_1$. Show that the resultant force acting on the surfaces $\theta = $ constant is zero, but that a couple of magnitude M acts in the 3-direction where

$$M = \tfrac{1}{2}W_0(r_1^2 - r_2^2) - \int_{r_1}^{r_2} rW\, \mathrm{d}r$$

and

$$W_0 = W(r_1) = W(r_2) = -K$$

Infinitesimal theory

In this chapter we show how the theory of finite deformation of an isotropic elastic solid can be reduced to a linear theory when the deformation is sufficiently small. This theory is commonly used in a variety of engineering applications.

4.1 Equations of motion

We now take $\mathbf{X}$ and t to be the independent variables, and we recall that in the material description the equations of linear and angular momentum may be written in the forms (1.14.9) and (1.14.10); that is

$$S_{Ki,K} + \rho_0 b_i = \rho_0 \frac{Dv_i}{Dt} \tag{4.1.1}$$

$$S_{Ki} x_{j,K} = S_{Kj} x_{i,K} \tag{4.1.2}$$

Using the displacement vector $\mathbf{u}$ defined in (1.7.1), and noting that

$$\frac{D\mathbf{u}}{Dt} = \frac{D\mathbf{x}}{Dt} = \mathbf{v} \tag{4.1.3}$$

we may write equations (4.1.1) and (4.1.2) in the form

$$S_{Ki,K} + \rho_0 b_i = \rho_0 \frac{D^2 u_i}{Dt^2} \tag{4.1.4}$$

$$S_{Ki}(\delta_{jK} + u_{j,K}) = S_{Kj}(\delta_{iK} + u_{i,K}) \tag{4.1.5}$$

Also the continuity equation (1.9.4) may be written as

$$\rho J = \rho \det(\delta_{iK} + u_{i,K}) = \rho_0 \tag{4.1.6}$$

In order to give meaning to the concept of small deformation, we define a measure of the magnitude of the displacement gradients

$$\varepsilon = (u_{i,K} u_{i,K})^{\frac{1}{2}} \tag{4.1.7}$$

Then, as $\varepsilon \to 0$, we see from (4.1.5) that

$$S_{Ki}\{\delta_{jK} + O(\varepsilon)\} = S_{Kj}\{\delta_{iK} + O(\varepsilon)\} \tag{4.1.8}$$

Also, recalling the definition (1.7.5),

$$E_{KL} = \tfrac{1}{2}(u_{K,L} + u_{L,K}) \tag{4.1.9}$$

we see that

$$J = \det(\delta_{iK} + u_{i,K}) = 1 + E_{KK} + O(\varepsilon^2) \tag{4.1.10}$$

Thus, (4.1.6) gives

$$\rho_0 = \rho\{1 + E_{KK} + O(\varepsilon^2)\} \tag{4.1.11}$$

If dV and dV_0 are corresponding volume elements in $\mathscr{C}_t$ and $\mathscr{C}_0$, respectively, the *volume strain* e is defined as

$$e = \frac{dV - dV_0}{dV_0} \tag{4.1.12}$$

and so, using (1.3.5),

$$e = J - 1 = E_{KK} + O(\varepsilon^2) \tag{4.1.13}$$

We also note here that the relation between the Cauchy stress and the Piola stress may be written, using (1.14.6),

$$\begin{aligned}
T_{ji} &= J^{-1}x_{j,K}S_{Ki} \\
&= \{1 - E_{KK} + O(\varepsilon^2)\}(\delta_{jK} + u_{j,K})S_{Ki} \\
&= S_{Ki}(\delta_{jK} + O(\varepsilon))
\end{aligned} \tag{4.1.14}$$

In order to discuss the boundary conditions, we recall that an element of surface dS_0 with unit normal $\mathbf{N}$ in $\mathscr{C}_0$ deforms into an element dS with unit normal $\mathbf{n}$ in $\mathscr{C}_t$, where using (1.14.4),

$$n_i x_{i,K}\, dS = JN_K\, dS_0$$

Hence,

$$n_i(\delta_{iK} + O(\varepsilon))\, dS = (1 + O(\varepsilon))N_K\, dS_0 \tag{4.1.15}$$

4.2   Stress–strain relations

We have shown that the constitutive equation for the stress may be written in the form (2.4.8)

$$S_{Ki} = x_{i,L}\left(\frac{\partial W}{\partial C_{KL}} + \frac{\partial W}{\partial C_{LK}}\right) \tag{4.2.1}$$

where, for an isotropic material, W is a function of the invariants of $\mathbf{C}$. However, in deriving the linear theory, it is more convenient to regard W as a function of the invariants of

$$\bar{\mathbf{E}} = \tfrac{1}{2}(\mathbf{C} - \mathbf{I}) \tag{4.2.2}$$

since $\bar{\mathbf{E}}$ vanishes when $\mathscr{C}_t$ coincides with $\mathscr{C}_0$, and from (1.7.4) we see that

$$\bar{\mathbf{E}} = O(\varepsilon) \tag{4.2.3}$$

Now one possible set of invariants of $\bar{\mathbf{E}}$ is

$$\operatorname{tr}\bar{\mathbf{E}} = \bar{E}_{KK}, \quad \operatorname{tr}(\bar{\mathbf{E}}^2) = \bar{E}_{KL}\bar{E}_{LK}, \quad \operatorname{tr}(\bar{\mathbf{E}}^3) = \bar{E}_{KL}\bar{E}_{LM}\bar{E}_{MK} \tag{4.2.4}$$

and we assume that, as $\varepsilon \to 0$, W has an asymptotic expansion of the form

$$W = \alpha\bar{E}_{KK} + \mu\bar{E}_{KL}\bar{E}_{KL} + \tfrac{1}{2}\lambda(\bar{E}_{KK})^2 + O(\varepsilon^3) \tag{4.2.5}$$

where α, μ, λ are constants. Since W depends only on the set of invariants (4.2.4), and in view of (4.2.3), the linear and quadratic terms in (4.2.5) are the only possible forms of this order. Using (4.2.1), (4.2.2), and (4.2.5), we find

$$\begin{aligned} S_{Ki} &= (\delta_{iL} + u_{i,L})\{\alpha\delta_{KL} + 2\mu\bar{E}_{KL} + \lambda\bar{E}_{MM}\delta_{KL} + O(\varepsilon^2)\} \\ &= \alpha(\delta_{iK} + u_{i,K}) + 2\mu\bar{E}_{iK} + \lambda\bar{E}_{MM}\delta_{iK} + O(\varepsilon^2) \end{aligned} \tag{4.2.6}$$

and so, when the body is in its reference configuration, $\mathbf{u} = \mathbf{0}$, $\bar{\mathbf{E}} = \mathbf{0}$, and

$$S_{Ki} = \alpha\delta_{iK}$$

We suppose, however, that the reference configuration is stress-free, and so we must take $\alpha = 0$. We also note from (1.7.4), (1.7.5), that

$$\bar{\mathbf{E}} = \mathbf{E} + O(\varepsilon^2)$$

and therefore (4.2.6) takes the form

$$S_{Ki} = 2\mu E_{iK} + \lambda E_{MM}\delta_{iK} + O(\varepsilon^2) \tag{4.2.7}$$

4.3 Formulation of the infinitesimal theory of elasticity

We are now in a position to state the equations governing the infinitesimal theory of elasticity. We take the basic equations as given in the previous two sections, but retain only the leading terms in the asymptotic expansions in powers of ε. We are using the material description, and so the independent variables appearing in the equations are the material coordinates X_A and the time t. The equations hold in the region $\mathscr{C}_0$ and are subject to boundary conditions on the surface S_0 of $\mathscr{C}_0$. As we no longer need to refer to partial derivatives with respect to the spatial coordinates, we now write all suffixes in upper-case letters. Thus the equations of motion (4.1.4) and (4.1.5) are written

$$S_{KL,K} + \rho_0 b_L = \rho_0 \frac{D^2 u_L}{Dt^2} \qquad (4.3.1)$$

$$S_{KL} = S_{LK} \qquad (4.3.2)$$

and the constitutive equation (4.2.7) becomes

$$S_{KL} = 2\mu E_{KL} + \lambda E_{MM} \delta_{KL} \qquad (4.3.3)$$

where

$$E_{KL} = \tfrac{1}{2}(u_{K,L} + u_{L,K}) \qquad (4.3.4)$$

The coefficients λ, μ are sometimes known as the *Lamé constants*.

It should perhaps be emphasised that these equations, governing the motion of the body, hold in the fixed region $\mathscr{C}_0$. Their solution gives us the displacement $\mathbf{u}(\mathbf{X}, t)$ as a vector field defined over $\mathscr{C}_0$ from which we can construct $\mathscr{C}_t$ since the points $\mathbf{x}$ of $\mathscr{C}_t$ are related to the points $\mathbf{X}$ of $\mathscr{C}_0$ by $\mathbf{x} = \mathbf{X} + \mathbf{u}$. Moreover, the stress tensor $\mathbf{S}(\mathbf{X}, t)$ is defined over $\mathscr{C}_0$ but describes the state of stress in $\mathscr{C}_t$. Neglecting the terms $O(\varepsilon)$ in (4.1.14), we see that $\mathbf{S} = \mathbf{T}$, and so the Piola stress $\mathbf{S}$ may be interpreted as the Cauchy stress $\mathbf{T}$ if required.

When the displacement $\mathbf{u}$ of the surface of $\mathscr{C}_t$ is specified, this is equivalent to giving $\mathbf{u}(\mathbf{X}, t)$ on S_0, and this boundary condition, together with the governing equations (4.3.1) to (4.3.4), define what is normally referred to as the *displacement boundary-value problem*.

Neglecting the terms $O(\varepsilon)$ in (4.1.15) we find

$$\mathbf{n} \, dS = \mathbf{N} \, dS_0$$

and, since $\mathbf{n}$ and $\mathbf{N}$ are unit vectors we must have

$$\mathbf{n} = \mathbf{N}, \qquad dS = dS_0$$

Hence, (1.14.5) and (1.14.11) imply

$$t_i = s_i = S_{Ki}N_K \tag{4.3.5}$$

and so, when the surface traction **t** is specified on the surface S of $\mathscr{C}_t$, this is equivalent to giving $S_{KL}N_K$ on the surface S_0 of $\mathscr{C}_0$. The set of governing differential equations, subject to boundary conditions of this kind, is normally referred to as the *traction boundary-value problem*.

Sometimes the displacement is specified on part of S_0 and the traction on the remainder. This we refer to as the *mixed boundary-value problem*.

For convenience later, we denote by $(\mathbf{u}, \mathbf{S})$ any solution of the governing equations (4.3.1) to (4.3.4). In view of the linearity of these equations, if $(\mathbf{u}_1, \mathbf{S}_1)$ and $(\mathbf{u}_2, \mathbf{S}_2)$ are solutions when $\mathbf{b} = \mathbf{0}$, then $(\mathbf{u}_1 + \mathbf{u}_2, \mathbf{S}_1 + \mathbf{S}_2)$ is also a solution. When $\mathbf{b} \neq \mathbf{0}$ a particular integral of (4.3.1) has to be found and this may then be added to any solution obtained with $\mathbf{b} = \mathbf{0}$.

In many of our applications we consider bodies in equilibrium. That is, we seek time-independent solutions $\mathbf{u}(\mathbf{X})$, $\mathbf{S}(\mathbf{X})$ of the *equilibrium equations*

$$S_{KL,K} + \rho_0 b_L = 0 \tag{4.3.6}$$

together with (4.3.2) to (4.3.4).

Example 4.3.1

Show that (4.3.3) may be solved for the strains in terms of the stresses in the form

$$2\mu E_{KL} = S_{KL} - \frac{\lambda}{3\lambda + 2\mu} S_{MM}\delta_{KL} \tag{4.3.7}$$

Example 4.3.2

Show that the strain-energy function may be expressed in the equivalent forms

$$W = \mu E_{KL}E_{KL} + \tfrac{1}{2}\lambda (E_{KK})^2 \tag{4.3.8}$$

$$4\mu W = S_{KL}S_{KL} - \frac{\lambda}{3\lambda + 2\mu} (S_{KK})^2 \tag{4.3.9}$$

4.4 Equation for the displacement vector

Substituting (4.3.4) into (4.3.3) and eliminating S_{KL} from (4.3.1), we obtain equations for the displacement components in the form

$$\mu(u_{K,LL} + u_{L,KL}) + \lambda u_{L,LK} + \rho_0 b_K = \rho_0 \frac{D^2 u_K}{Dt^2} \qquad (4.4.1)$$

That is,

$$(\lambda + \mu)u_{L,KL} + \mu u_{K,LL} + \rho_0 b_K = \rho_0 \frac{D^2 u_K}{Dt^2} \qquad (4.4.2)$$

which, in vector notation, may be written

$$(\lambda + \mu)\,\text{grad div }\mathbf{u} + \mu\,\nabla^2\mathbf{u} + \rho_0\mathbf{b} = \rho_0 \frac{D^2\mathbf{u}}{Dt^2} \qquad (4.4.3)$$

Alternatively, using the relation

$$\text{curl curl }\mathbf{u} = \text{grad div }\mathbf{u} - \nabla^2\mathbf{u}$$

we can write

$$(\lambda + 2\mu)\,\text{grad div }\mathbf{u} - \mu\,\text{curl curl }\mathbf{u} + \rho_0\mathbf{b} = \rho_0 \frac{D^2\mathbf{u}}{Dt^2} \qquad (4.4.4)$$

Of course, in these equations the vector operators grad, div, and curl are defined relative to the material coordinates X_K.

Example 4.4.1

(i) Show that

$$(\mathbf{N} \times \text{curl }\mathbf{u})_K = N_L u_{L,K} - N_L u_{K,L} \qquad (4.4.5)$$

(ii) Use (1.14.11), (4.3.3), (4.3.4), and (4.4.5) to show that

$$\mathbf{s}(\mathbf{N}) = \lambda\mathbf{N}\,\text{div }\mathbf{u} + 2\mu(\mathbf{N} \cdot \nabla)\mathbf{u} + \mu\mathbf{N} \times \text{curl }\mathbf{u} \qquad (4.4.6)$$

4.5 Compatibility equations for the components of the infinitesimal strain tensor

Given a displacement field $\mathbf{u}$, the components of the infinitesimal strain tensor,

$$E_{KL} = \tfrac{1}{2}(u_{K,L} + u_{L,K}) \qquad (4.5.1)$$

are easily calculated. However, a displacement field **u** satisfying (4.5.1) for a given symmetric tensor **E** may not exist. We now consider what conditions must be satisfied by the components E_{KL} to ensure the existence of displacement components satisfying (4.5.1). We first recall a fundamental theorem of vector analysis.

Theorem.   If **u** is a differentiable vector field, and

$$\varepsilon_{KLM}u_{M,L} = 0 \tag{4.5.2}$$

in a simply connected region R, then there exists a scalar field ϕ such that

$$u_K = \phi_{,K} \tag{4.5.3}$$

in R.

Example 4.5.1

Let A be a fixed point of R and let P be a point in R with coordinates X_K. Show that

$$\phi = \int_A^P \mathbf{u} \cdot d\mathbf{X} \tag{4.5.4}$$

where the line integral is taken along any curve in R, has the required property (4.5.3).

Example 4.5.2

If **A** is a differentiable tensor field and

$$\varepsilon_{KLM}A_{NM,L} = 0 \tag{4.5.5}$$

in a simply connected region R, show that there exists a vector field **u** such that

$$A_{NM} = u_{N,M} \tag{4.5.6}$$

in R.

Let **a** be any constant vector, and let **p** be the vector with components

$$p_K = a_L A_{LK}$$

Then

$$\varepsilon_{LMN}p_{N,M} = 0$$

from (4.5.5). Hence, from the above theorem, a scalar ϕ exists such that

$$p_K = \phi_{,K}$$

and using (4.5.4) we may take ϕ in the form

$$\phi = \int_A^P \mathbf{p} \cdot d\mathbf{X} = a_K \int_A^P A_{KL} \, dX_L$$

Now define

$$u_K = \int_A^P A_{KL} \, dX_L$$

Then

$$\phi = a_K u_K$$

and

$$a_K A_{KL} = p_L = \phi_{,L} = a_K u_{K,L} \tag{4.5.7}$$

Since (4.5.7) must hold for all vectors $\mathbf{a}$, we must have

$$A_{KL} = u_{K,L} \tag{4.5.8}$$

Example 4.5.3

If $\mathbf{A}$ is a differentiable tensor field such that

$$\varepsilon_{KLM} A_{NM,L} = 0, \qquad A_{KK} = 0 \tag{4.5.9}$$

in a simply connected region R, show that there exists a skew-symmetric tensor $\mathbf{W}$ such that

$$A_{KL} = \varepsilon_{KMN} W_{LN,M} \tag{4.5.10}$$

In view of Example 4.5.2, there exists a vector $\mathbf{u}$ such that

$$A_{KL} = u_{K,L} \tag{4.5.11}$$

Now let

$$W_{LN} = \varepsilon_{LNK} u_K \tag{4.5.12}$$

Then

$$
\begin{aligned}
\varepsilon_{KMN} W_{LN,M} &= \varepsilon_{KMN} \varepsilon_{LNP} u_{P,M} \\
&= (\delta_{KP}\delta_{ML} - \delta_{KL}\delta_{MP}) u_{P,M} \\
&= u_{K,L} - \delta_{KL} u_{M,M} \\
&= u_{K,L} \\
&= A_{KL}
\end{aligned}
$$

where use has been made of $(4.5.9)_2$, $(4.5.11)$ and the $\varepsilon - \delta$ formula of tensor analysis.

We are now in a position to discuss conditions on the components E_{KL} which are necessary and sufficient for the existence of displacement components u_K satisfying $(4.5.1)$. We first define $\boldsymbol{\omega} = \text{curl } \mathbf{u}$, so that

$$\omega_K = \varepsilon_{KLM} u_{M,L} \tag{4.5.13}$$

Then

$$2\varepsilon_{KLM} E_{PM,L} = \varepsilon_{KLM}(u_{P,ML} + u_{M,PL})$$
$$= \varepsilon_{KLM} u_{M,PL}$$
$$= \omega_{K,P} \tag{4.5.14}$$

Hence,

$$2\varepsilon_{QSP}\varepsilon_{KLM} E_{PM,LS} = \varepsilon_{QSP}\omega_{K,PS} = 0 \tag{4.5.15}$$

Equations $(4.5.15)$ consist of six relations which must necessarily be satisfied by the components E_{KL}. Inspection of $(4.5.15)$ reveals that they fall into two groups of three equations, the equations in each group having similar form. The first group may be obtained by taking the suffixes Q and K to be equal, and a typical member of the group which results when $Q = K = 1$ is

$$E_{33,22} + E_{22,33} - 2E_{23,23} = 0 \tag{4.5.16}$$

The second group is obtained by taking $Q \neq K$, a typical member when $Q = 2$, $K = 1$ being

$$E_{13,23} + E_{23,13} - E_{12,33} - E_{33,12} = 0 \tag{4.5.17}$$

Equations $(4.5.15)$ are known as the *infinitesimal strain compatibility equations*. To show that they are sufficient to ensure the existence of displacement components u_K in a simply connected region R, we first define

$$A_{KL} = \varepsilon_{KPQ} E_{LQ,P} \tag{4.5.18}$$

Then

$$\varepsilon_{KLM} A_{NM,L} = 0$$

from $(4.5.15)$. Also,

$$A_{KK} = \varepsilon_{KPQ} E_{KQ,P} = 0$$

from the symmetry of E_{KQ}. Hence, using Example 4.5.3, there exists a skew-symmetric tensor **W** such that

$$A_{KL} = \varepsilon_{KPQ} W_{LQ,P} \tag{4.5.19}$$

Using (4.5.18) and (4.5.19), we see that

$$\varepsilon_{KPQ}(E_{LQ,P} - W_{LQ,P}) = 0 \tag{4.5.20}$$

and so, using Example 4.5.2, we see that a vector **u** exists such that

$$E_{LQ} - W_{LQ} = u_{L,Q} \tag{4.5.21}$$

Now **E** is symmetric and **W** is skew-symmetric. Hence

$$E_{LQ} = \tfrac{1}{2}(u_{L,Q} + u_{Q,L})$$

The conditions (4.5.15) are therefore sufficient for the existence of a displacement field **u** from which **E** may be derived using (4.5.1).

4.6   Energy equations and uniqueness of solution

When a boundary-value problem is formulated, natural questions to ask are: does a solution exist? And, if so, is it possible to find more than one solution? Although a discussion of the existence of solutions is beyond the scope of this volume, we do consider here the question of uniqueness of solution.

In most of our applications of the theory, we choose situations in which solutions may be obtained by inverse methods; that is, we postulate certain forms of the displacement and stress fields, and then show that these fields satisfy the equations (4.3.1) to (4.3.4) together with prescribed boundary conditions. Having found a solution, it is then natural to question whether another solution of the equations could be found satisfying the same boundary conditions. For physically realistic materials, we may answer in the negative; the solution is unique. The proof of this statement, however, depends on the form of the strain-energy function W, and the energy equation in our linear theory.

We consider first the displacement and stress fields for a body in equilibrium; that is, we consider stress components S_{KL} and displacement components u_K satisfying the equilibrium equations

$$S_{KL,K} + \rho_0 b_L = 0 \tag{4.6.1}$$

and the stress–strain equations

$$S_{KL} = 2\mu E_{KL} + \lambda E_{MM}\delta_{KL} \tag{4.6.2}$$

where

$$E_{KL} = \tfrac{1}{2}(u_{K,L} + u_{L,K}) \tag{4.6.3}$$

Taking the scalar product of (4.6.1) with u_L, and integrating over an arbitrary volume V_0, bounded by a surface S_0 in $\mathscr{C}_0$, we have

$$\int_{V_0} u_L S_{KL,K}\, dV_0 + \int_{V_0} \rho_0 b_L u_L\, dV_0 = 0 \tag{4.6.4}$$

Now (4.6.4) may be written

$$\int_{V_0} \frac{\partial}{\partial X_K}(u_L S_{KL})\, dV_0 + \int_{V_0} \rho_0 b_L u_L\, dV_0 = \int_{V_0} u_{L,K} S_{KL}\, dV_0$$

or, using the divergence theorem, (4.3.5), (4.3.2), and (4.3.4), we have

$$\int_{S_0} u_L s_L\, dS_0 + \int_{V_0} \rho_0 b_L u_L\, dV_0 = \int_{V_0} S_{KL} E_{KL}\, dV_0 \tag{4.6.5}$$

The left-hand side of (4.6.5) may be interpreted as the sum of work done by the surface and body forces in the infinitésimal displacement **u**, and the right-hand side denotes the work done by the stresses throughout V_0. Using (4.6.2) and (4.3.8), the right-hand side of (4.6.5) may be written

$$\int_{V_0} S_{KL} E_{KL}\, dV_0 = \int_{V_0} \{\lambda(E_{KK})^2 + 2\mu E_{KL} E_{KL}\}\, dV_0$$

$$= 2\int_{V_0} W\, dV_0 \tag{4.6.6}$$

Example 4.6.1

Defining the *strain deviator* $\mathbf{E}'$ by

$$E'_{KL} = E_{KL} - \tfrac{1}{3}E_{MM}\delta_{KL} \tag{4.6.7}$$

show that $E'_{KK} = 0$, and

$$2W = \kappa(E_{KK})^2 + 2\mu E'_{KL} E'_{KL} \tag{4.6.8}$$

where $\kappa = \lambda + \frac{2}{3}\mu$

We now assume that

$$\kappa > 0, \ \mu > 0 \qquad (4.6.9)$$

and we see later that these inequalities are necessary for physically reasonable response of the material (see Section 4.7). They are also of fundamental importance in the proof of the following uniqueness theorems.

Theorem.   There is at most one solution $(\mathbf{u}, \mathbf{S})$ of the equations (4.6.1) to (4.6.3) which satisfies the boundary conditions

$$\mathbf{u} = \mathbf{u}_0 \quad \text{on} \quad S_u, \qquad \mathbf{s}(\mathbf{N}) = \mathbf{s}_0 \quad \text{on} \quad S_s \qquad (4.6.10)$$

where the complete boundary of the body is $S_0 = S_u + S_s$, $(S_u \neq 0)$, and $\mathbf{u}_0$ and $\mathbf{s}_0$ are vector fields specified on S_u and S_s, respectively. If $S_u = 0$, any two solutions differ by at most a rigid-body displacement.

Proof.   Suppose that two displacement fields $\mathbf{u}_1$ and $\mathbf{u}_2$ exist which satisfy $(4.6.10)_1$ on S_u and have associated stress fields which satisfy $(4.6.10)_2$ on S_s and $(4.6.1)$ throughout the region V_0 occupied by the body in $\mathscr{C}_0$. Then the displacement field $\mathbf{u} = \mathbf{u}_1 - \mathbf{u}_2$ has the boundary value $\mathbf{u} = \mathbf{0}$ on S_u, and has an associated stress field which gives $\mathbf{s}(\mathbf{N}) = \mathbf{0}$ on S_s and satisfies the equilibrium equations $(4.6.1)$ with $\mathbf{b} = \mathbf{0}$. Hence, using equations $(4.6.5)$, $(4.6.6)$, $(4.6.8)$, we see that the strain components E_{KL} derived from $\mathbf{u}$ must satisfy

$$\int_{V_0} \{\kappa (E_{KK})^2 + 2\mu E'_{KL} E'_{KL}\} \, dV_0 = 0 \qquad (4.6.11)$$

In view of the inequalities $(4.6.9)$, and the fact that $(E_{KK})^2$ and $E'_{KL} E'_{KL}$ are non-negative, equation $(4.6.11)$ implies

$$\int_{V_0} (E_{KK})^2 \, dV_0 = 0, \qquad \int_{V_0} E'_{KL} E'_{KL} \, dV_0 = 0$$

from which it follows that

$$E_{KK} = 0, \qquad E'_{KL} = 0$$

Hence,

$$E_{KL} = 0 \qquad (4.6.12)$$

Equation (4.6.12) means that **u** corresponds to a rigid-body displacement (see Example 1.7.1). However, since **u** = **0** on S_u, and provided S_u exists, we must have **u** = **0** everywhere in the body. This means that $\mathbf{u}_1 = \mathbf{u}_2$, which proves the uniqueness of the solution. If S_u does not exist, so that surface traction is specified over the whole boundary, then $\mathbf{u}_1$ and $\mathbf{u}_2$ may differ by a rigid-body displacement.

Consider now the dynamic problem. In this case, the displacement and stress components must satisfy the equations of motion

$$S_{KL,K} + \rho_0 b_L = \rho_0 \frac{D^2 u_L}{Dt^2} \tag{4.6.13}$$

and the stress–strain relations

$$S_{KL} = 2\mu E_{KL} + \lambda E_{MM} \delta_{KL} \tag{4.6.14}$$

where

$$E_{KL} = \tfrac{1}{2}(u_{K,L} + u_{L,K}) \tag{4.6.15}$$

Example 4.6.2

Taking the scalar product of (4.6.13) with Du_L/Dt and integrating over an arbitrary volume V_0, show that

$$\int_{S_0} s_L \frac{Du_L}{Dt} \, dS_0 + \int_{V_0} \rho_0 b_L \frac{Du_L}{Dt} \, dV_0 = \int_{V_0} S_{KL} \frac{DE_{KL}}{Dt} \, dV_0$$

$$+ \frac{d}{dt} \int_{V_0} \tfrac{1}{2} \rho_0 \frac{Du_L}{Dt} \frac{Du_L}{Dt} \, dV_0 \tag{4.6.16}$$

Example 4.6.3

Using (4.6.2) and (4.6.6), show that

$$\int_{V_0} S_{KL} \frac{DE_{KL}}{Dt} \, dV_0 = \frac{d}{dt} \int_{V_0} W \, dV_0 \tag{4.6.17}$$

Theorem. There is at most one solution (**u**, **S**) of the equations (4.6.13) to (4.6.15) which satisfies the boundary conditions

$$\mathbf{u} = \mathbf{u}_0 \quad \text{on} \quad S_u, \qquad \mathbf{s(N)} = \mathbf{s}_0 \quad \text{on} \quad S_s \tag{4.6.18}$$

and the initial conditions

$$\mathbf{u} = \mathbf{0}, \qquad \frac{D\mathbf{u}}{Dt} = \mathbf{v}_0 \qquad \text{at time } t = 0 \qquad (4.6.19)$$

where, at any instant, the complete boundary of the body is $S_u + S_s$, $\mathbf{u}_0$, and $\mathbf{s}_0$ are vector fields specified on S_u and S_s respectively, and $\mathbf{v}_0$ is a vector field specified over the whole body at time $t = 0$.

Proof. As in the proof of the previous theorem, we suppose that two solutions exist satisfying equations (4.6.13) to (4.6.15) and the boundary and initial conditions (4.6.18), (4.6.19). If the displacements are denoted by $\mathbf{u}_1$ and $\mathbf{u}_2$, then the displacement $\mathbf{u} = \mathbf{u}_1 - \mathbf{u}_2$ satisfies (4.6.13) with $\mathbf{b} = \mathbf{0}$, together with boundary conditions

$$\mathbf{u} = \mathbf{0} \quad \text{on} \quad S_u, \qquad \mathbf{s}(\mathbf{N}) = \mathbf{0} \quad \text{on} \quad S_s \qquad (4.6.20)$$

and initial conditions

$$\mathbf{u} = \mathbf{0}, \qquad \frac{D\mathbf{u}}{Dt} = \mathbf{0} \qquad \text{at } t = 0 \qquad (4.6.21)$$

Also, since $\mathbf{u} = \mathbf{0}$ on S_u for all t, $D\mathbf{u}/Dt = 0$ on S_u. Thus, applying (4.6.16) to the displacement field $\mathbf{u}$ and the associated stress field $\mathbf{S}$, and using (4.6.17), (4.6.20), we find

$$\frac{d}{dt} \left\{ \int_{V_0} W \, dV_0 + \int_{V_0} \tfrac{1}{2}\rho_0 \frac{Du_L}{Dt} \frac{Du_L}{Dt} \, dV_0 \right\} = 0 \qquad (4.6.22)$$

where V_0 is the region occupied by the body in $\mathscr{C}_0$.

Now in view of (4.6.21), initially we must have $D\mathbf{u}/Dt = \mathbf{0}$, and since $\mathbf{u} = \mathbf{0}$ then $\mathbf{E} = \mathbf{0}$ and $W = 0$. Hence,

$$\int_{V_0} W \, dV_0 + \int_{V_0} \tfrac{1}{2}\rho_0 \frac{Du_L}{Dt} \frac{Du_L}{Dt} \, dV_0 = 0$$

Since each integrand is non-negative, we must have

$$\int_{V_0} W \, dV_0 = 0, \qquad \int_{V_0} \tfrac{1}{2}\rho_0 \frac{Du_L}{Dt} \frac{Du_L}{Dt} \, dV_0 = 0$$

and therefore $W = 0$, $D\mathbf{u}/Dt = \mathbf{0}$. Initially $\mathbf{u} = \mathbf{0}$, and so $\mathbf{u} = \mathbf{0}$ for all time. This proves that $\mathbf{u}_1 = \mathbf{u}_2$.

4.7   Pure homogeneous deformations

We consider here some solutions of the equations of linear elasticity corresponding to the pure homogeneous deformations discussed in Sections 3.4 to 3.7 for finite deformation. Now using (1.7.1) the deformation (3.4.1) may be described by the displacement field

$$u_1 = \alpha_1 X_1, \quad u_2 = \alpha_2 X_2, \quad u_3 = \alpha_3 X_3 \qquad (4.7.1)$$

where the constants α_K are related to the constants λ_K by

$$\alpha_K = \lambda_K - 1 \qquad (4.7.2)$$

Using (4.3.4), we find that the infinitesimal strain tensor has components

$$\mathbf{E} = \begin{pmatrix} \alpha_1 & 0 & 0 \\ 0 & \alpha_2 & 0 \\ 0 & 0 & \alpha_3 \end{pmatrix} \qquad (4.7.3)$$

and, using the stress–strain relations (4.3.3) we see that the stress tensor is given by

$$\mathbf{S} = \lambda(\alpha_1 + \alpha_2 + \alpha_3)\begin{pmatrix} 1 & 0 & 0 \\ 0 & 1 & 0 \\ 0 & 0 & 1 \end{pmatrix} + 2\mu\begin{pmatrix} \alpha_1 & 0 & 0 \\ 0 & \alpha_2 & 0 \\ 0 & 0 & \alpha_3 \end{pmatrix} \qquad (4.7.4)$$

Since all the stress components are constant, the equilibrium equations (4.3.6) are satisfied identically when the body force is absent, and so (4.7.1) describes a possible static displacement field. Some special choices of the constants α_K lead to important uniform stress systems.

(i) Purely normal stress

The simplest choice is to take all the α_K equal. Then from (4.7.4) we have

$$\mathbf{S} = -p\mathbf{I} \qquad (4.7.5)$$

where p is a constant given by

$$\alpha_1 = \alpha_2 = \alpha_3 = -p/(3\lambda + 2\mu) \qquad (4.7.6)$$

This is the state of stress that we have called purely normal stress (see Section 1.11). If $p > 0$, the stress state is one of hydrostatic pressure, and the deformation (4.7.1) is a uniform contraction.

Using (4.1.13) we see that the volume strain is

$$e = E_{KK} = \alpha_1 + \alpha_2 + \alpha_3 = -p/\kappa \qquad (4.7.7)$$

where $\kappa = \lambda + \frac{2}{3}\mu$. The constant κ is known as the *bulk modulus*. A stress system of this kind arises in a body of arbitrary shape when the surface is subjected to a uniform hydrostatic pressure p. Since we expect a hydrostatic pressure to decrease the volume of the material, we assume

$$\kappa > 0 \qquad (4.7.8)$$

When $\alpha_1 = \alpha_2 = \alpha_3 > 0$, we refer to the deformation (4.7.1) as a uniform expansion.

(ii) Uniaxial tension

We observe from (4.7.4) that the stress components S_{11} and S_{22} are zero when

$$\alpha_1 = \alpha_2 = -\nu\alpha_3 \qquad (4.7.9)$$

where

$$\nu = \tfrac{1}{2}\lambda/(\lambda + \mu) \qquad (4.7.10)$$

and then the only non-vanishing stress component is

$$S_{33} = E\alpha_3 = EE_{33} \qquad (4.7.11)$$

where

$$E = \mu(3\lambda + 2\mu)/(\lambda + \mu) \qquad (4.7.12)$$

As we noted in Section 1.11, if $S_{33} > 0$ such a stress system is called a uniaxial tension in the 3-direction. The magnitude of the stress S_{33} is proportional to the strain $E_{33} = \alpha_3$, and the constant of proportionality E is known as *Young's modulus*. From (4.7.9) it is also clear that the strains $E_{11} = \alpha_1$ and $E_{22} = \alpha_2$ are also proportional to E_{33}. The constant ν, which is the ratio of the lateral contraction $-E_{11}$ to the axial extension E_{33}, is called *Poisson's ratio*.

As an example in which such a stress system arises, consider a bar of length l and square cross-section with side of length a, and suppose that the coordinate axes are taken as shown in Fig. 4.1. Suppose that a uniform normal stress S is applied to the ends while the remaining surfaces are left free of applied traction.

Fig. 4.1

Then the boundary conditions are:

$$\text{On} \quad X_3 = l, \quad 0 \leq X_1 \leq a, \quad 0 \leq X_2 \leq a; \quad \mathbf{s}(\mathbf{e}_3) = S\mathbf{e}_3,$$
$$\text{i.e.} \quad S_{33} = S, \quad S_{31} = S_{32} = 0.$$
$$\text{On} \quad X_3 = 0, \quad 0 \leq X_1 \leq a, \quad 0 \leq X_2 \leq a; \quad \mathbf{s}(-\mathbf{e}_3) = -S\mathbf{e}_3,$$
$$\text{i.e.} \quad S_{33} = S, \quad S_{31} = S_{32} = 0.$$
$$\text{On} \quad X_1 = 0, a, \quad 0 \leq X_2 \leq a, \quad 0 \leq X_3 \leq l;$$
$$S_{11} = S_{12} = S_{13} = 0.$$
$$\text{On} \quad X_2 = 0, a, \quad 0 \leq X_1 \leq a, \quad 0 \leq X_3 \leq l;$$
$$S_{22} = S_{21} = S_{23} = 0.$$

$$(4.7.13)$$

Clearly therefore, the displacement field (4.7.1), subject to (4.7.9) and with $\alpha_3 = S/E$, gives rise to stress components which satisfy the equilibrium equations, in the absence of body force, and the boundary conditions (4.7.13). This displacement field may be written

$$u_1 = -\nu S X_1 / E, \quad u_2 = -\nu S X_2 / E, \quad u_3 = S X_3 / E \quad (4.7.14)$$

Now after deformation the surfaces $X_3 = 0$ and $X_3 = l$ become

$$x_3 = [X_3 + u_3]_{X_3=0} = 0, \quad \text{and} \quad x_3 = [X_3 + u_3]_{X_3=l} = l + Sl/E$$

and so the increase in the length of the bar is Sl/E. Likewise, the surfaces $X_1 = 0$ and $X_1 = a$ become

$$x_1 = [X_1 + u_1]_{X_1=0} = 0, \quad \text{and} \quad x_1 = [X_1 + u_1]_{X_1=a} = a - \nu Sa/E$$

and so the decrease in width is $\nu Sa/E$.

In all known elastic materials, a uniaxial tension produces an extension in the direction of the tension and so we assume that

$$E > 0 \quad\quad\quad (4.7.15)$$

Associated with such an extension we would also expect a lateral contraction, and so we also assume that

$$\nu > 0 \quad\quad\quad (4.7.16)$$

Example 4.7.1

A cylindrical bar of length l and with generators parallel to the X_3-axis has arbitrary cross-section. The generating surface is free of applied traction, and a uniform normal stress S is applied to the end sections. Using (4.3.5), show that the boundary condition on the generating surface is satisfied if

$$S_{11}N_1 + S_{21}N_2 = 0$$
$$S_{12}N_1 + S_{22}N_2 = 0$$
$$S_{13}N_1 + S_{23}N_2 = 0$$

where $(N_1, N_2, 0)$ denotes the normal to the generating surface. Hence, show that the displacement components (4.7.14) satisfy the conditions of this problem also.

(iii) Pure shear

Earlier in Section 3.7 we discussed the simple shear deformation

$$x_1 = X_1 + \kappa X_2, \quad x_2 = X_2, \quad x_3 = X_3 \qquad (4.7.17)$$

For this deformation the displacement components are

$$u_1 = \kappa X_2, \quad u_2 = 0, \quad u_3 = 0 \qquad (4.7.18)$$

and so all the infinitesimal strain components are zero except

$$E_{12} = \tfrac{1}{2}\kappa \qquad (4.7.19)$$

The only non-vanishing stress is therefore

$$S_{12} = \mu\kappa \qquad (4.7.20)$$

Such a stress system which clearly satisfies the equilibrium equations (4.3.6) in the absence of body force is known as pure shear, and the constant of proportionality μ in (4.7.20) is sometimes referred to as the *shear modulus*.

In the deformation of a cuboid by surface shear tractions, as illustrated in Fig. 4.2, the angle of shear θ is equal to κ, to the first order in κ (see Section 1.8), and as θ would be positive for positive S_{12} in all known materials, we assume that

$$\mu > 0 \qquad (4.7.21)$$

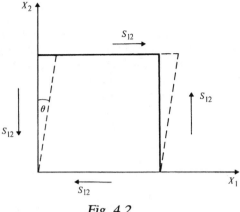

Fig. 4.2

Example 4.7.2

Show that, in terms of Young's modulus E and Poisson's ratio ν, the constitutive equation (4.3.3) may be written

$$S_{KL} = \frac{E\nu}{(1+\nu)(1-2\nu)} E_{MM}\delta_{KL} + \frac{E}{1+\nu} E_{KL} \qquad (4.7.22)$$

4.8   Values of the elastic constants

In the last section we have shown that, for physically reasonable behaviour, the elastic constants should satisfy the inequalities (4.7.8), (4.7.15), and (4.7.21) namely,

$$\kappa > 0, \quad E > 0, \quad \mu > 0 \qquad (4.8.1)$$

and we are also assuming (4.7.16),

$$\nu > 0 \qquad (4.8.2)$$

Recalling the definitions

$$\kappa = \tfrac{1}{3}(3\lambda + 2\mu), \quad E = \mu\frac{(3\lambda + 2\mu)}{\lambda + \mu}, \quad \nu = \frac{\lambda}{2(\lambda + \mu)} \qquad (4.8.3)$$

we see that the inequalities (4.8.1) and (4.8.2) imply

$$\lambda + \mu > 0, \quad \lambda > 0, \quad 0 < \nu < \tfrac{1}{2} \qquad (4.8.4)$$

Throughout this text we assume that these inequalities are satisfied.

   Table 4.1 gives the numerical values of the elastic constants for some common metals at 20 °C.

Table 4.1

	E unit $= 10^{10}$N/m^2	κ unit $= 10^{10}$N/m^2	ν
Aluminium	7.03	7.55	0.345
Copper	12.98	13.78	0.343
Soft iron	12.14	16.98	0.293
Mild steel	21.19	16.92	0.291

Example 4.8.1

Calculate the increase in length and decrease in diameter when a cylindrical copper bar of length 300 mm, and diameter 10 mm is subjected to a uniaxial tension along its length by the application of forces of magnitude 100 N distributed uniformly over the ends.

Example 4.8.2

Show that $\mu = 3E\kappa/(9\kappa - E)$.

Find the angle of shear when mild steel is subjected to a pure shear stress of 3×10^7 N/m^2.

4.9   Spherical symmetry

Considerable simplification of the basic equations of elasticity results when the problem exhibits a high degree of symmetry. In this section we consider problems in which the displacement field is directed radially from a fixed point, which we take to be the origin, and has magnitude depending only on the distance R from that fixed point. We suppose also that the body force is similarly restricted. In other words, if $\mathbf{e}_R$ denotes the unit vector in the radial direction,

$$\mathbf{u} = u(R)\mathbf{e}_R, \qquad \mathbf{b} = b(R)\mathbf{e}_R \qquad (4.9.1)$$

Hence,

$$\text{div } \mathbf{u} = \text{div}\left(\frac{u}{R}\mathbf{X}\right) = \frac{u}{R}\text{div }\mathbf{X} + \mathbf{X} \cdot \text{grad}\left(\frac{u}{R}\right)$$

$$= 3\frac{u}{R} + \frac{du}{dR} - \frac{u}{R}$$

$$= \frac{1}{R^2}\frac{d}{dR}(R^2 u) \qquad (4.9.2)$$

Also

$$\text{curl } \mathbf{u} = \text{curl} \left(\frac{u}{R} \mathbf{X} \right) = \left(\text{grad } \frac{u}{R} \right) \times \mathbf{X} + \frac{u}{R} \text{ curl } \mathbf{X}$$
$$= \mathbf{0} \qquad (4.9.3)$$

since $\text{grad }(u/R)$ must be in the $\mathbf{e}_R$-direction, and $\text{curl } \mathbf{X} = \mathbf{0}$. The relations (4.9.2) and (4.9.3) may also be written down directly using the formulae for div and curl in spherical polar coordinates.

Substituting (4.9.2) and (4.9.3) into the equation of equilibrium

$$(\lambda + 2\mu) \text{ grad div } \mathbf{u} - \mu \text{ curl curl } \mathbf{u} + \rho_0 \mathbf{b} = \mathbf{0} \qquad (4.9.4)$$

which is obtained from (4.4.4), we find

$$(\lambda + 2\mu) \frac{d}{dR} \left\{ \frac{1}{R^2} \frac{d}{dR} (R^2 u) \right\} + \rho_0 b = 0 \qquad (4.9.5)$$

When $b = 0$, (4.9.5) has the general solution

$$u = AR + B/R^2 \qquad (4.9.6)$$

where A and B are constants of integration. These are normally determined by the boundary conditions. In order to discuss traction boundary conditions, we now evaluate $\mathbf{s}(\mathbf{e}_R)$. Since the components of $\mathbf{e}_R$ are X_K/R, using (4.3.5), we find

$$s_K = S_{LK} X_L / R \qquad (4.9.7)$$

Using (4.9.6), the components of the displacement $(4.9.1)_1$ are

$$u_K = \left(AR + \frac{B}{R^2} \right) \frac{X_K}{R} = A X_K + B \frac{X_K}{R^3} \qquad (4.9.8)$$

Now

$$R^2 = X_K X_K$$

and so

$$\frac{\partial R}{\partial X_K} = \frac{X_K}{R} \qquad (4.9.9)$$

Hence,

$$u_{K,L} = \left(A + \frac{B}{R^3} \right) \delta_{KL} - 3B \frac{X_K X_L}{R^5} \qquad (4.9.10)$$

and
$$u_{K,K} = 3A \tag{4.9.11}$$

The stress–strain relations (4.3.3) now give the stress components in the form

$$S_{KL} = 3A\lambda\delta_{KL} + 2\mu\left(A + \frac{B}{R^3}\right)\delta_{KL} - 6\mu B\frac{X_K X_L}{R^5} \tag{4.9.12}$$

Using (4.9.7) we find that $\mathbf{s}(\mathbf{e_R})$ has components

$$s_K = (3\lambda + 2\mu)A\frac{X_K}{R} + 2\mu B\frac{X_K}{R^4} - 6\mu B\frac{X_K}{R^4}$$

$$= \left\{(3\lambda + 2\mu)A - 4\mu\frac{B}{R^3}\right\}\frac{X_K}{R} \tag{4.9.13}$$

Since X_K/R are the components of $\mathbf{e_R}$, and using $(4.8.3)_1$, we find

$$\mathbf{s}(\mathbf{e_R}) = \left\{3\kappa A - 4\mu\frac{B}{R^3}\right\}\mathbf{e_R} \tag{4.9.14}$$

Thus the surface traction on surfaces $R = $ constant is entirely normal. Using (4.9.6) or (4.9.14), A and B may be evaluated from suitable boundary conditions.

Example 4.9.1

A pressure vessel is in the form of a shell bounded by concentric spheres $R = a$ and $R = b$ $(a < b)$. The inner surface is subjected to a pressure p_1 and the outer surface to a pressure p_2. Find the displacement field in the shell, and also the normal stress acting tangentially to the surfaces $R = a$ and $R = b$.

The boundary conditions are

$$\mathbf{s}(-\mathbf{e_R}) = p_1\mathbf{e_R} \qquad \text{on } R = a$$
$$\mathbf{s}(\mathbf{e_R}) = -p_2\mathbf{e_R} \qquad \text{on } R = b$$

Using (4.9.14) the equations for A and B are therefore

$$3\kappa A - 4\mu B/a^3 = -p_1$$
$$3\kappa A - 4\mu B/b^3 = -p_2$$

giving

$$A = \frac{p_1 a^3 - p_2 b^3}{3\kappa(b^3 - a^3)}, \qquad B = \frac{(p_1 - p_2)a^3 b^3}{4\mu(b^3 - a^3)} \tag{4.9.15}$$

Substituting (4.9.15) into (4.9.6) and (4.9.1)$_1$ gives the displacement field in the shell.

To find the normal stress acting tangentially to the surfaces of the shell, sometimes called the peripheral stress, we use spherical polar coordinates (R, Θ, Φ) and the notation of Section 1.15, remembering that $\mathbf{T}$ may be identified with $\mathbf{S}$ to our present order of approximation. From symmetry,

$$S_{\Phi\Phi} = S_{\Theta\Theta} \qquad (4.9.16)$$

and these stress components measure the peripheral stress on $R = a$ and $R = b$.

Now T_{kk} is an invariant (see Section 1.12) and so in our infinitesimal theory S_{KK} is an invariant. Hence, we may deduce from the stress–strain relations (4.3.3)

$$S_{RR} + S_{\Theta\Theta} + S_{\Phi\Phi} = S_{KK} = 3\kappa E_{KK} = 3\kappa u_{K,K} \qquad (4.9.17)$$

On $R = a$ and $R = b$, S_{RR} is equal to $-p_1$ and $-p_2$ respectively. Also we see from (4.9.11) that $u_{K,K}$ has the value $3A$ at all points of the material. Therefore, on $R = a$,

$$S_{\Theta\Theta} = S_{\Phi\Phi} = \tfrac{1}{2}(9\kappa A + p_1) = \frac{2p_1 a^3 + (p_1 - 3p_2)b^3}{2(b^3 - a^3)},$$

and on $R = b$,

$$S_{\Theta\Theta} = S_{\Phi\Phi} = \tfrac{1}{2}(9\kappa A + p_2) = \frac{(3p_1 - p_2)a^3 - 2p_2 b^3}{2(b^3 - a^3)}.$$

The peripheral stress on the surface of the shell is an important factor in determining the onset of fracture.

4.10   The Boussinesq–Papkovitch–Neuber solution

When the elastic stress and displacement fields do not have a sufficiently high degree of symmetry for the governing equations to reduce to ordinary differential equations, the most common approach to three-dimensional equilibrium problems involves the use of particular solutions of the equilibrium equations which may be expressed in terms of harmonic functions.

When an elastic body is in equilibrium, in the absence of body force, we see from (4.4.3) and (4.7.10) that the displacement $\mathbf{u}$

must satisfy

$$\text{grad div } \mathbf{u} + (1 - 2\nu) \nabla^2 \mathbf{u} = \mathbf{0} \qquad (4.10.1)$$

We consider two particular solutions of this equation.

Example 4.10.1

If

$$\mathbf{u}_1 = \text{grad } \phi \qquad (4.10.2)$$

where ϕ is any function satisfying $\nabla^2 \phi = 0$, show that $\mathbf{u}_1$ is a solution of (4.10.1).

Clearly

$$\text{div } \mathbf{u}_1 = \nabla^2 \phi = 0$$

and

$$\nabla^2 \mathbf{u}_1 = \text{grad } \nabla^2 \phi = \mathbf{0}$$

Hence, (4.10.1) is satisfied.

Example 4.10.2

If

$$\mathbf{u}_2 = 4(1 - \nu)\boldsymbol{\psi} - \text{grad } (\mathbf{X} \cdot \boldsymbol{\psi}) \qquad (4.10.3)$$

where $\boldsymbol{\psi}$ is any vector field satisfying $\nabla^2 \boldsymbol{\psi} = \mathbf{0}$, show that $\mathbf{u}_2$ is a solution of (4.10.1).

Using the suffix notation

$$(X_K \psi_K)_{,L} = \psi_L + X_K \psi_{K,L}$$

Therefore,

$$u_{2L} = (3 - 4\nu)\psi_L - X_K \psi_{K,L}$$

and

$$u_{2L,L} = (3 - 4\nu)\psi_{L,L} - X_K \psi_{K,LL} - \psi_{L,L}$$
$$= 2(1 - 2\nu)\psi_{L,L},$$

since $\psi_{K,LL} = 0$. Hence,

$$\text{div } \mathbf{u}_2 = 2(1 - 2\nu) \text{ div } \boldsymbol{\psi} \qquad (4.10.4)$$

Also

$$u_{2L,KK} = (3 - 4\nu)\psi_{L,KK} - (\psi_{K,L} + X_M\psi_{M,LK})_{,K}$$
$$= -2\psi_{K,LK}$$

Hence,

$$\nabla^2 \mathbf{u}_2 = -2 \operatorname{grad} \operatorname{div} \boldsymbol{\psi} \tag{4.10.5}$$

Using (4.10.4) and (4.10.5), we see that $\mathbf{u}_2$ satisfies (4.10.1).

The sum of these two displacement fields $\mathbf{u}_1 + \mathbf{u}_2$ is known as the Boussinesq–Papkovitch–Neuber solution; and it can be shown that, under certain smoothness assumptions and subject to conditions on the elastic constants weaker than (4.8.1) and (4.8.2), there exist a scalar field ϕ and vector field $\boldsymbol{\psi}$ such that any solution of (4.10.1) may be expressed in the form $\mathbf{u} = \mathbf{u}_1 + \mathbf{u}_2$ (Gurtin (1972), p. 140).

Moreover, if the body is bounded by a surface of revolution about the Z-axis, and the boundary conditions are such that the displacement components in cylindrical polar coordinates (R, Θ, Z) (see Section 1.15),

$$X_1 = R \cos \Theta, \quad X_2 = R \sin \Theta$$

take the form

$$u_R = u_R(R, Z), \quad u_\Theta = 0, \quad u_Z = u_Z(R, Z) \tag{4.10.6}$$

then there exist harmonic scalar functions $\phi(R, Z)$ and $\psi(R, Z)$ such that any solution of the form (4.10.6) may be expressed as $\mathbf{u} = \mathbf{u}_1 + \mathbf{u}_2$ with $\boldsymbol{\psi} = \psi \mathbf{e}_Z$. We make use of this form of solution in some problems discussed later.

For the solution of particular problems in which we do not need to make use of the repeated-suffix summation convention, we shall from here onwards use (X, Y, Z) in place of (X_1, X_2, X_3) to denote the components of the position vector $\mathbf{X}$. Likewise, the stress components S_{11}, S_{12}, etc., will be denoted by S_{XX}, S_{XY}, etc., the unit vectors $\mathbf{e}_1$, $\mathbf{e}_2$, $\mathbf{e}_3$ by $\mathbf{e}_X$, $\mathbf{e}_Y$, $\mathbf{e}_Z$, and $\mathbf{u}$ by (u, v, w).

Example 4.10.3

For every differentiable function $f(X, Y, Z)$, homogeneous of degree n in (X, Y, Z), show that

$$X \frac{\partial f}{\partial X} + Y \frac{\partial f}{\partial Y} + Z \frac{\partial f}{\partial Z} = nf \tag{4.10.7}$$

The function $f(X, Y, Z)$ is defined to be homogeneous of degree n in (X, Y, Z) if for all α $(\alpha \neq 0$ if $n < 0)$,

$$f(\alpha X, \alpha Y, \alpha Z) = \alpha^n f(X, Y, Z) \qquad (4.10.8)$$

Differentiating (4.10.8) with respect to α and then putting $\alpha = 1$ gives (4.10.7).

Example 4.10.4

If $\mathbf{e}_R$ denotes the unit vector in the direction of $\mathbf{X}$, show that when ϕ and ψ are homogeneous functions of degree n in X, Y, Z, the stress vectors $\mathbf{s}(\mathbf{e}_R)$ corresponding to the displacements $\mathbf{u}_1$ and $\mathbf{u}_2$, with $\boldsymbol{\psi} = \psi \mathbf{e}_Z$ are given by

$$R\mathbf{s}(\mathbf{e}_R) = 2(n-1)\mu \, \text{grad} \, \phi \qquad (4.10.9)$$

$$R\mathbf{s}(\mathbf{e}_R) = 4\mu\nu\mathbf{X}\frac{\partial \psi}{\partial Z} + 2\mu(1 - 2\nu)n\psi\mathbf{e}_Z \qquad (4.10.10)$$

$$+ 2\mu(2 - 2\nu - n)Z \, \text{grad} \, \psi$$

respectively.

Recalling that div $\mathbf{u}_1 = 0$ and noting that curl $\mathbf{u}_1 = \mathbf{0}$, we see from (4.4.6) that for any unit normal $\mathbf{N}$

$$\mathbf{s}(\mathbf{N}) = 2\mu(\mathbf{N} \cdot \nabla)\mathbf{u}_1 \qquad (4.10.11)$$

Hence,

$$R\mathbf{s}(\mathbf{e}_R) = 2\mu(\mathbf{X} \cdot \nabla)\mathbf{u}_1 \qquad (4.10.12)$$

Since ϕ is homogeneous of degree n in (X, Y, Z), $\mathbf{u}_1 = \text{grad} \, \phi$ is homogeneous of degree $n - 1$ in (X, Y, Z), and so, using (4.10.7), (4.10.12) gives

$$R\mathbf{s}(\mathbf{e}_R) = 2(n-1)\mu \, \text{grad} \, \phi$$

Using (4.10.4) in (4.4.6) we see that, corresponding to $\mathbf{u}_2$,

$$R\mathbf{s}(\mathbf{e}_R) = 4\mu\nu\mathbf{X}\frac{\partial \psi}{\partial Z} + 2\mu(\mathbf{X} \cdot \nabla)\{4(1 - \nu)\psi\mathbf{e}_Z - \text{grad} \, (Z\psi)\}$$

$$+ \mu\mathbf{X} \times \text{curl} \, \{4(1 - \nu)\psi\mathbf{e}_Z\}$$

since, using (4.7.10), we have $2\lambda(1 - 2\nu) = 4\mu\nu$.

Hence,

$$\begin{aligned}
R\mathbf{s}(\mathbf{e}_R) &= 4\mu\nu\mathbf{X}\frac{\partial\psi}{\partial Z} + 8\mu(1-\nu)(\mathbf{X}\cdot\boldsymbol{\nabla})\psi\mathbf{e}_Z \\
&\quad - 2\mu(\mathbf{X}\cdot\boldsymbol{\nabla})(\psi\mathbf{e}_Z + Z\,\mathrm{grad}\,\psi) \\
&\quad + 4\mu(1-\nu)\{Z\,\mathrm{grad}\,\psi - (\mathbf{X}\cdot\boldsymbol{\nabla})\psi\mathbf{e}_Z\} \\
&= 4\mu\nu\mathbf{X}\frac{\partial\psi}{\partial Z} + 8\mu(1-\nu)(\mathbf{X}\cdot\boldsymbol{\nabla})\psi\mathbf{e}_Z \\
&\quad - 2\mu(\mathbf{X}\cdot\boldsymbol{\nabla})\psi\mathbf{e}_Z - 2\mu Z\,\mathrm{grad}\,\psi \\
&\quad - 2\mu Z(\mathbf{X}\cdot\boldsymbol{\nabla})\,\mathrm{grad}\,\psi + 4\mu(1-\nu)Z\,\mathrm{grad}\,\psi \\
&\quad - 4\mu(1-\nu)(\mathbf{X}\cdot\boldsymbol{\nabla})\psi\mathbf{e}_Z \\
&= 4\mu\nu\mathbf{X}\frac{\partial\psi}{\partial Z} + 2\mu(1-2\nu)(\mathbf{X}\cdot\boldsymbol{\nabla})\psi\mathbf{e}_Z \\
&\quad + 2\mu(1-2\nu)Z\,\mathrm{grad}\,\psi - 2\mu Z(\mathbf{X}\cdot\boldsymbol{\nabla})\,\mathrm{grad}\,\psi
\end{aligned}$$

Again making use of (4.10.7) and the fact that ψ is homogeneous of degree n in (X, Y, Z), we find

$$\begin{aligned}
R\mathbf{s}(\mathbf{e}_R) &= 4\mu\nu\mathbf{X}\frac{\partial\psi}{\partial Z} + 2\mu(1-2\nu)n\psi\mathbf{e}_Z \\
&\quad + 2\mu(1-2\nu)Z\,\mathrm{grad}\,\psi - 2\mu Z(n-1)\,\mathrm{grad}\,\psi \\
&= 4\mu\nu\mathbf{X}\frac{\partial\psi}{\partial Z} + 2\mu(1-2\nu)n\psi\mathbf{e}_Z \\
&\quad + 2\mu(2-2\nu-n)Z\,\mathrm{grad}\,\psi.
\end{aligned}$$

Example 4.10.5

Show that the values of $\mathbf{s}(\mathbf{e}_Z)$, corresponding to the displacement fields described in Example 4.10.4 are given by

$$\mathbf{s}(\mathbf{e}_Z) = 2\mu\,\mathrm{grad}\,\frac{\partial\phi}{\partial Z} \tag{4.10.13}$$

and

$$\mathbf{s}(\mathbf{e}_Z) = 2\mu\left\{\frac{\partial\psi}{\partial Z}\mathbf{e}_Z + (1-2\nu)\,\mathrm{grad}\,\psi - Z\,\mathrm{grad}\,\frac{\partial\psi}{\partial Z}\right\} \tag{4.10.14}$$

respectively.

4.11 Concentrated loads

In many practical situations, the region over which force is applied to an elastic body may be regarded as small compared with the overall dimensions of the body. The force may usually be thought of as being distributed over a small surface or volume element. In such cases, it is often convenient to represent the effect of such applied forces by a point singularity in the displacement and stress fields.

Here we examine the basic singular displacement fields corresponding to isolated forces acting at a point within an infinite elastic medium, and also at a point on the plane boundary of an elastic half-space. In the first case we take a displacement field with a singularity at the origin, and calculate the force resultant on a spherical surface centred at the origin. We show that this resultant is independent of the radius of the surface, and we assume that an isolated force acts at the origin which balances this resultant. However, our analysis does not show that the singularity is the same as would be obtained in the limit of a sequence of solutions in which a given applied force is distributed uniformly over an element of volume of a size which tends to zero. Likewise, in the case of the isolated force acting on a half-space, we assume that the singularity corresponds to an isolated force which balances the resultant force on a hemisphere centred at the origin. A discussion of the precise conditions under which singularities may be obtained from limiting applied-force distributions is beyond the scope of this volume. We refer the interested reader to Gurtin (1972), Sections 51 to 53.

4.12 Isolated point force in an infinite medium

Consider the displacement field $\mathbf{u}_1 + \mathbf{u}_2$ given by (4.10.2) and (4.10.3) with

$$\phi = 0, \quad \mathbf{\psi} = \psi \mathbf{e}_Z, \quad \psi = A/R \qquad (4.12.1)$$

where A is a constant, and $R^2 = X^2 + Y^2 + Z^2$. Since ψ is a homogeneous harmonic function of degree -1 in (X, Y, Z), we may use the result (4.10.10) to show that

$$Rs(\mathbf{e_R}) = 4\mu\nu\frac{\partial\psi}{\partial Z}\mathbf{X} - 2\mu(1-2\nu)\psi\mathbf{e_Z} + 2\mu(3-2\nu)Z\,\mathrm{grad}\,\psi$$

$$= -A\left\{6\mu\frac{Z\mathbf{X}}{R^3} + 2\mu(1-2\nu)\frac{\mathbf{e_Z}}{R}\right\} \qquad (4.12.2)$$

Consider now a spherical surface S in the material, with centre at the origin O, and arbitrary radius ε. The resultant force, acting on the inside surface of S, is

$$\int_S \mathbf{s}(-\mathbf{e_R})\,\mathrm{d}S = -\int_S \mathbf{s}(\mathbf{e_R})\,\mathrm{d}S$$

$$= 2\mu\frac{A}{\varepsilon}\int_S \left\{\frac{3Z\mathbf{X}}{\varepsilon^3} + (1-2\nu)\frac{\mathbf{e_Z}}{\varepsilon}\right\}\mathrm{d}S$$

Since the surface S is positioned symmetrically with respect to the coordinate planes we have

$$\int_S XZ\,\mathrm{d}S = \int_S YZ\,\mathrm{d}S = 0$$

and

$$\int_S X^2\,\mathrm{d}S = \int_S Y^2\,\mathrm{d}S = \int_S Z^2\,\mathrm{d}S$$

$$= \tfrac{1}{3}\int_S (X^2+Y^2+Z^2)\,\mathrm{d}S$$

$$= \tfrac{1}{3}\int_S \varepsilon^2\,\mathrm{d}S$$

Hence, the resultant force is

$$2\mu A\int_S \left\{\frac{\mathbf{e_Z}}{\varepsilon^2} + (1-2\nu)\frac{\mathbf{e_Z}}{\varepsilon^2}\right\}\mathrm{d}S = 2\mu A\frac{2(1-\nu)}{\varepsilon^2}4\pi\varepsilon^2\mathbf{e_Z}$$

$$= 16\mu A(1-\nu)\pi\mathbf{e_Z} \qquad (4.12.3)$$

This force is clearly independent of ε. Thus, in the limit as $\varepsilon \to 0$, the resultant force remains constant. We may therefore regard the above solution as that corresponding to an infinite medium from which an arbitrarily small sphere of material has been removed about O, and to which certain surface tractions have been applied. These tractions on the spherical surface have a resultant

$P\mathbf{e}_Z$ provided we choose

$$A = P/16\mu\pi(1-\nu) \tag{4.12.4}$$

We emphasise that other displacement singularities can have the same properties, and, as mentioned in Section 4.11, the precise sense in which the solution determined by (4.12.1) represents an isolated force cannot be discussed further here.

Using (4.10.2), (4.10.3), (4.12.1), and (4.12.4), the explicit form for the displacement components may be written

$$\mathbf{u} = \frac{P}{16\mu\pi(1-\nu)} \left(\frac{XZ}{R^3}, \frac{YZ}{R^3}, \frac{Z^2}{R^3} + \frac{3-4\nu}{R} \right) \tag{4.12.5}$$

Of course, as $R \to 0$, the displacement gradients become unbounded and the infinitesimal theory becomes invalid.

4.13   Isolated point force on a plane boundary

We suppose that elastic material occupies the half-space $Z \geqslant 0$, and that, apart from the origin, the surface $Z = 0$ is free of applied traction. We consider again a displacement field $\mathbf{u}_1 + \mathbf{u}_2$ given by (4.10.2), (4.10.3) with a singularity at the origin, but in this case we take

$$\phi = \log(R+Z), \quad \boldsymbol{\psi} = \psi\mathbf{e}_Z, \quad \psi = \frac{1}{R} \tag{4.13.1}$$

where, as before, $R^2 = X^2 + Y^2 + Z^2$. Using the formula (4.10.13), we see that the components of $\mathbf{s}(\mathbf{e}_Z)$ corresponding to $\mathbf{u}_1$ are

$$S_{XZ} = -2\mu\frac{X}{R^3}, \quad S_{YZ} = -2\mu\frac{Y}{R^3}, \quad S_{ZZ} = -2\mu\frac{Z}{R^3} \tag{4.13.2}$$

Using (4.10.14) we find the components of $\mathbf{s}(\mathbf{e}_Z)$ corresponding to $\mathbf{u}_2$ to be

$$S_{XZ} = 2\mu\left\{ (1-2\nu)\frac{\partial\psi}{\partial X} - Z\frac{\partial^2\psi}{\partial X\,\partial Z} \right\} = -2\mu\left\{ (1-2\nu)\frac{X}{R^3} + \frac{3XZ^2}{R^5} \right\}$$

$$S_{YZ} = 2\mu\left\{ (1-2\nu)\frac{\partial\psi}{\partial Y} - Z\frac{\partial^2\psi}{\partial Y\,\partial Z} \right\} = -2\mu\left\{ (1-2\nu)\frac{Y}{R^3} + \frac{3YZ^2}{R^5} \right\}$$

$$S_{ZZ} = 2\mu\left\{ 2(1-\nu)\frac{\partial\psi}{\partial Z} - Z\frac{\partial^2\psi}{\partial Z^2} \right\} = -2\mu\left\{ (1-2\nu)\frac{Z}{R^3} + \frac{3Z^3}{R^5} \right\}$$

$$\tag{4.13.3}$$

The form of the stress components (4.13.2) and (4.13.3) shows that the corresponding stress components associated with the displacement field

$$\mathbf{u} = A\{\mathbf{u}_1 - (1 - 2\nu)\mathbf{u}_2\} \tag{4.13.4}$$

where A is a constant, are

$$S_{XZ} = -6\mu A X Z^2 / R^5, \quad S_{YZ} = -6\mu A Y Z^2 / R^5, \quad S_{ZZ} = -6\mu A Z^3 / R^5 \tag{4.13.5}$$

which clearly leave the surface $Z = 0$ (excluding the origin) free of traction.

Consider now a hemisphere S of radius ε, centred at the origin, and lying in the material of the half-space $Z \geqslant 0$. The resultant force acting on the inside surface of S must, from symmetry, act in the Z-direction, and using (4.3.5), its magnitude is

$$
\begin{aligned}
\int_S \mathbf{s}(-\mathbf{e}_R) \cdot \mathbf{e}_Z \, dS &= -\int_S \mathbf{s}(\mathbf{e}_R) \cdot \mathbf{e}_Z \, dS \\
&= -\int_S \varepsilon^{-1}(S_{XZ}X + S_{YZ}Y + S_{ZZ}Z) \, dS \\
&= 6\mu A \varepsilon^{-6} \int_S Z^2(X^2 + Y^2 + Z^2) \, dS \\
&= 6\mu A \varepsilon^{-4} \int_S Z^2 \, dS \\
&= 6\mu A \int_0^{\pi/2} 2\pi \cos^2 \theta \sin \theta \, d\theta \\
&= 12\mu A \pi [-\tfrac{1}{3} \cos^3 \theta]_0^{\pi/2} \\
&= 4\mu\pi A \tag{4.13.6}
\end{aligned}
$$

Here again, this resultant force is independent of ε, and so in the limit as $\varepsilon \to 0$ the resultant force remains constant. Hence, the displacement field (4.13.4) may be regarded as corresponding to the semi-infinite region $Z \geqslant 0$ from which an arbitrarily small hemisphere has been removed, the resultant force on the hemisphere being $P\mathbf{e}_Z$, provided we choose

$$A = P/4\mu\pi \tag{4.13.7}$$

Example 4.13.1

Show that the displacement components in the above solution are

$$\frac{P}{4\mu\pi}\left(\frac{ZX}{R^3}-\frac{(1-2\nu)X}{R(R+Z)}, \quad \frac{ZY}{R^3}-\frac{(1-2\nu)Y}{R(R+Z)}, \quad \frac{Z^2}{R^3}+\frac{2(1-\nu)}{R}\right)$$

Examples 4

1. An elastic circular cylinder of finite length fits accurately into a rigid, smooth cylindrical tube with axis lying along the Z-axis, and with open ends. Body force is assumed to be absent, and equal and opposite compressive normal stresses are applied uniformly over the end faces of the cylinder. By considering the displacement field (4.7.1) with appropriate choices of α_i, show that

$$S_{ZZ}=(\lambda+2\mu)E_{ZZ}$$

and determine the surface traction on the curved surface of the cylinder.

Does this displacement field satisfy the corresponding conditions for an elastic cylinder of arbitrary cross-section?

2. Elastic material extending to infinity in all directions contains a rigid spherical inclusion of radius a. If the material is in equilibrium under the action of a uniform pressure p at infinity, and body force is absent, show that the displacement throughout the material is given by

$$\mathbf{u}=\frac{-p}{3\lambda+2\mu}\left(R-\frac{a^3}{R^2}\right)\mathbf{e_R}$$

where R denotes the radial distance from the centre of the sphere.

3. Elastic material with Lamé constants λ_1, μ_1 occupies the spherical region $R < a$, and elastic material with Lamé constants λ_2, μ_2 occupies the region $a \leqslant R \leqslant b$. When the surface $R = b$ is subjected to a uniform pressure p, and body force is absent, explain why the stress and displacement vectors are continuous across $R = a$ and find the equilibrium displacement in both regions of the body.

4. Elastic material with Lamé constants λ, μ, and density ρ, occupies the spherical region $R \leqslant a$. The surface $R = a$ is free of applied traction, but a body force of magnitude kR per unit mass acts towards the centre of the sphere, where k is a constant. Show that the equilibrium displacement field in the sphere is given by

$$\mathbf{u} = \frac{\rho k}{10(\lambda + 2\mu)} \left\{ R^3 - \frac{5\lambda + 6\mu}{3\lambda + 2\mu} a^2 R \right\} \mathbf{e}_R$$

Find the deformed configuration of the surface.

5. When $\mathbf{u} = u(R)\mathbf{e}_R$ and $\mathbf{b} = b(R)\mathbf{e}_R$, where R denotes the first *cylindrical* polar coordinate $(R^2 = X^2 + Y^2)$, show that the equilibrium equations reduce to

$$(\lambda + 2\mu) \frac{d}{dR} \left\{ \frac{1}{R} \frac{d}{dR} (Ru) \right\} + \rho b(R) = 0$$

When body force is absent, show that u has the form $u = AR + B/R$, where A and B are constants, and that

$$\mathbf{s}(\mathbf{e}_R) = \left\{ 2(\lambda + \mu)A - 2\mu \frac{B}{R^2} \right\} \mathbf{e}_R$$

Elastic material occupies the region $a \leqslant R \leqslant b$. The surface $R = a$ is fixed and the surface $R = b$ is subjected to a uniform pressure p. Show that the displacement in the material is given by

$$u = \frac{-pb^2}{2(\lambda + \mu)b^2 + 2\mu a^2} \left\{ R - \frac{a^2}{R} \right\}$$

6. Show that when the harmonic functions ϕ and ψ take the values

$$\phi = \frac{AZ}{R^3}, \qquad \psi = \frac{B}{R}$$

where A, B are constants, and $R^2 = X^2 + Y^2 + Z^2$, the elastic displacement fields $\mathbf{u}_1 = \operatorname{grad} \phi$, $\mathbf{u}_2 = 4(1 - \nu)\psi\mathbf{e}_Z - \operatorname{grad}(Z\psi)$ reduce to

$$\mathbf{u}_1 = \frac{A}{R^3} \mathbf{e}_Z - \frac{3AZ}{R^4} \mathbf{e}_R, \qquad \mathbf{u}_2 = (3 - 4\nu) \frac{B}{R} \mathbf{e}_Z + \frac{BZ}{R^2} \mathbf{e}_R$$

A rigid sphere $R \leqslant a$ is embedded in an infinite elastic material, and adheres to the material at all points of its surface. The sphere is given a displacement $\alpha\mathbf{e}_Z$ while the material is held fixed at infinity. Show that the displacement field $\mathbf{u}_1 + \mathbf{u}_2$ satisfies the boundary conditions provided that

$$A = \frac{\alpha a^3}{2(5-6\nu)}, \qquad B = \frac{3\alpha a}{2(5-6\nu)}$$

Show that the displacement on the Z-axis has magnitude

$$\frac{\alpha a}{(5-6\nu)Z^3}\{6(1-\nu)Z^2 - a^2\}$$

7. If the Z-axis is taken to be vertically downwards, and the only non-zero stress component in a material under the influence of gravity is S_{ZZ}, show that the equilibrium equations are satisfied provided that

$$S_{ZZ} = -\rho_0 g Z + \text{constant}$$

The generators of a heavy uniform cylinder of length l and arbitrary cross-section are vertical, the upper end and cylindrical surface being free of applied traction. Show that the stress components associated with the displacement field

$$\mathbf{u} = \left(\frac{\rho_0 g \nu}{E}XZ, \ \frac{\rho_0 g \nu}{E}YZ, \ -\frac{\rho_0 g}{2E}Z^2 - \frac{\rho_0 g \nu}{2E}(X^2+Y^2)\right)$$

satisfy the equilibrium equations and the boundary conditions.

Describe two ways in which the cylinder could be supported at the lower end while maintaining this displacement field.

If the cross-section of the undeformed cylinder is the circular region $X^2 + Y^2 \leq a^2$, draw sketches to illustrate the deformed configuration.

8. Show that the displacement field $\mathbf{u} = \text{curl } \boldsymbol{\Psi}$ satisfies the equilibrium equations, in the absence of body force, provided that $\nabla^2 \boldsymbol{\Psi} = \mathbf{0}$.

A rigid sphere of radius a is embedded in an infinite elastic material to which it adheres at all points of its boundary. The material is held fixed at infinity while the sphere is given an infinitesimal rotation ω about the Z-axis which passes through the centre of the sphere. Show that the boundary conditions are satisfied by a displacement field of the above type when $\boldsymbol{\Psi} = a^3 \omega \mathbf{e}_Z/R$, $R^2 = X^2 + Y^2 + Z^2$. Show that on $R = a$,

$$\mathbf{s}(\mathbf{e}_R) = 3\mu\frac{\omega}{a}\mathbf{X} \times \mathbf{e}_Z$$

and that the couple which must be applied to the sphere to cause the rotation has magnitude $8\pi a^3 \mu \omega$.

5

Anti-plane strain, plane strain, and generalised plane stress

Here we consider displacement fields which are independent of the Z-coordinate. The bodies in which these fields arise would normally be cylinders with generators parallel to the Z-axis, under the action of prescribed surface tractions or displacements which are also independent of Z. We show that such displacement fields may be regarded as the superposition of two displacement fields, each governed by independent sets of equations; the first being a displacement field in the Z-direction (anti-plane strain) and the second being parallel to the XY-plane (plane strain). With a small modification, the plane-strain analysis is also applicable to the stretching of thin plates (generalised plane stress).

5.1 Basic equations

We suppose that the cylindrical body meets the XY-plane in a region R bounded by a curve C, and we denote by S the surface of the cylinder. Since the stress and displacement components are independent of Z, the stress–strain relations (4.3.3) reduce to

$$\left. \begin{array}{ll} S_{XX} = (\lambda + 2\mu)\dfrac{\partial u}{\partial X} + \lambda\dfrac{\partial v}{\partial Y}, & S_{YY} = \lambda\dfrac{\partial u}{\partial X} + (\lambda + 2\mu)\dfrac{\partial v}{\partial Y} \\[3mm] S_{XY} = \mu\left(\dfrac{\partial u}{\partial Y} + \dfrac{\partial v}{\partial X}\right), & S_{ZZ} = \lambda\left(\dfrac{\partial u}{\partial X} + \dfrac{\partial v}{\partial Y}\right) \end{array} \right\} \quad (5.1.1)$$

$$S_{XZ} = \mu\frac{\partial w}{\partial X}, \quad S_{YZ} = \mu\frac{\partial w}{\partial Y} \qquad (5.1.2)$$

The equilibrium equations (4.3.6) in the absence of body force reduce to

$$\left.\begin{aligned}\frac{\partial S_{XX}}{\partial X}+\frac{\partial S_{XY}}{\partial Y}=0\\[1mm]\frac{\partial S_{XY}}{\partial X}+\frac{\partial S_{YY}}{\partial Y}=0\end{aligned}\right\}\tag{5.1.3}$$

$$\frac{\partial S_{XZ}}{\partial X}+\frac{\partial S_{YZ}}{\partial Y}=0.\tag{5.1.4}$$

The above equations clearly fall into two distinct groups which may be solved separately. Equations (5.1.1) and (5.1.3) constitute a set of six equations for the quantities S_{XX}, S_{YY}, S_{XY}, S_{ZZ}, u, and v; equations (5.1.2) and (5.1.4) constitute a set of three equations for S_{XZ}, S_{YZ}, and w. A body in which all the variables S_{XZ}, S_{YZ}, and w are identically zero and for which the set S_{XX}, S_{YY}, S_{XY}, S_{ZZ}, u, and v satisfies (5.1.1) and (5.1.3), is said to be in a state of *plane strain*. A body in which S_{XX}, S_{YY}, S_{XY}, S_{ZZ}, u, and v are all identically zero and the set S_{XZ}, S_{YZ}, and w satisfies (5.1.2) and (5.1.4) is said to be in a state of *anti-plane strain*.

5.2 Anti-plane strain

We consider first the simplest of the above states, namely that of anti-plane strain governed by equations (5.1.2) and (5.1.4). Substituting equations (5.1.2) into (5.1.4), we find

$$\frac{\partial^2 w}{\partial X^2}+\frac{\partial^2 w}{\partial Y^2}=0\tag{5.2.1}$$

In other words, w must be a plane harmonic function in R. Problems in which the displacement w is specified on the boundary C of the cross-section of the body reduce to the well-known Dirichlet problem of potential theory.

The surface traction on the cylindrical surface of the body has components $s_K = S_{LK}N_L$, so that

$$s_X = s_Y = 0,\qquad s_Z = \mu\left(l\frac{\partial w}{\partial X}+m\frac{\partial w}{\partial Y}\right)\tag{5.2.2}$$

where $(l, m, 0)$ are the components of the outward normal $\mathbf{N}$ to S. Hence, the surface traction is in the Z-direction and has magnitude $\mu(\mathbf{N} \cdot \nabla)w$. The traction boundary-value problem therefore reduces to the Neumann problem of potential theory.

Example 5.2.1

Elastic material occupies the region $a \leqslant R \leqslant b$, $R^2 = X^2 + Y^2$. The surface $R = a$ is held fixed, while the surface $R = b$ is displaced a distance ε in the Z-direction. Assuming that body force is absent, find the displacement and stress components in the material.

In view of the symmetry of the body, and boundary conditions, about the Z-axis, the displacement may be assumed to depend only on R. In this case, equation (5.2.1) becomes

$$\frac{1}{R}\frac{d}{dR}\left(R\frac{dw}{dR}\right) = 0$$

which has solution

$$w = A \log R + B$$

where A and B are constants. The boundary conditions are

$$w = 0 \quad \text{on} \quad R = a, \qquad w = \varepsilon \quad \text{on} \quad R = b$$

Hence

$$A \log a + B = 0, \quad A \log b + B = \varepsilon$$

and so

$$A = \varepsilon/\log(b/a), \quad B = -\varepsilon \log a/\log(b/a)$$

Thus

$$w = \varepsilon \log(R/a)/\log(b/a)$$

Referred to cylindrical polar coordinates (R, Θ, Z) and using the notation of Section 1.15, remembering that $\mathbf{T}$ may be identified with $\mathbf{S}$ to our order of approximation, the only non-vanishing stress component is S_{RZ}, which using (5.2.2), is given by

$$S_{RZ} = \frac{\mu}{R}\left(X\frac{\partial w}{\partial X} + Y\frac{\partial w}{\partial Y}\right)$$

$$= \mu\frac{dw}{dR} = \frac{\varepsilon}{R \log(b/a)}$$

5.3   Plane strain. Equations for the stress field

As mentioned in Section 5.1, plane strain involves the stress components S_{XX}, S_{XY}, S_{YY}, S_{ZZ} and the displacement components u and v. However, we note that S_{ZZ} occurs only in $(5.1.1)_4$ and so the main task is the solution of $(5.1.1)_{1,2,3}$ together with $(5.1.3)$ for the variables S_{XX}, S_{XY}, S_{YY}, u, and v. Having found these, S_{ZZ} may be evaluated directly from $(5.1.1)_4$ and gives the normal stress in the Z-direction which must be applied in order to maintain the deformation.

We show here that the governing equations take on a simple form when the displacement components u and v are eliminated. The equations then depend only on the stress components S_{XX}, S_{XY}, and S_{YY}, and have a particularly elegant solution which we discuss in the next sections.

If surface tractions are specified on the boundary of the body and only the stress field within the body is required, the problem may be posed entirely in terms of the stresses; no reference need be made to the displacement. If, however, the displacement components are required, these may most easily be found by the complex-variable methods discussed in Section 5.6.

Instead of eliminating u and v directly from $(5.1.1)$, we recall that, in a simply connected region R, the compatibility equations $(4.5.15)$ for E_{KL} are necessary and sufficient for the existence of a displacement field. Now in plane strain the only non-vanishing strain components are E_{XX}, E_{XY}, and E_{YY} and these depend only on X and Y. Hence, the only compatibility equation which is not satisfied identically is

$$E_{XX,YY} + E_{YY,XX} - 2E_{XY,XY} = 0 \qquad (5.3.1)$$

Using $(5.1.1)$, $(4.3.7)$, and $(4.7.10)$, the strains may be written in terms of the stresses:

$$2\mu E_{XX} = S_{XX} - \nu\hat{\Theta}, \quad 2\mu E_{YY} = S_{YY} - \nu\hat{\Theta}, \quad 2\mu E_{XY} = S_{XY} \qquad (5.3.2)$$

where

$$\hat{\Theta} = S_{XX} + S_{YY} \qquad (5.3.3)$$

Also

$$S_{ZZ} = \nu\hat{\Theta} \qquad (5.3.4)$$

Hence, the compatibility equation $(5.3.1)$ may be written in terms of the stresses as

$$\frac{\partial^2}{\partial Y^2}(S_{XX} - \nu\hat{\Theta}) + \frac{\partial^2}{\partial X^2}(S_{YY} - \nu\hat{\Theta}) - 2\frac{\partial^2 S_{XY}}{\partial X \partial Y} = 0 \qquad (5.3.5)$$

Example 5.3.1

Deduce equation (5.3.5) directly from equations (5.3.2) by writing the strain components in terms of the displacement components u and v, and eliminating u and v.

Using (5.1.3), (5.3.5) may be written in the compact form

$$(1 - \nu)\, \nabla^2\hat{\Theta} = 0 \qquad (5.3.6)$$

where $\nabla^2 \equiv \partial^2/\partial X^2 + \partial^2/\partial Y^2$ is the two-dimensional Laplacian operator. Finally, therefore, using (4.7.16) and (4.8.4), we may say that for the simply connected region R, the governing equations for the plane-strain problem in terms of the stresses are

$$\frac{\partial S_{XX}}{\partial X} + \frac{\partial S_{XY}}{\partial Y} = \frac{\partial S_{XY}}{\partial X} + \frac{\partial S_{YY}}{\partial Y} = 0, \quad \nabla^2\hat{\Theta} = 0 \qquad (5.3.7)$$

Stress components satisfying these equations have corresponding strain components which satisfy the strain compatibility equations and so therefore an associated displacement field exists.

5.4   The Airy stress function

We already know that a differentiable vector field $\mathbf{F}$ is irrotational, that is curl $\mathbf{F} = \mathbf{0}$, if and only if there exists a scalar field Ω such that $\mathbf{F} = \text{grad}\,\Omega$ (cf. Section 4.5). In the particular case when $\mathbf{F} = (-g(X, Y), f(X, Y), 0)$, it follows from this result that

$$\frac{\partial f}{\partial X} + \frac{\partial g}{\partial Y} = 0 \qquad (5.4.1)$$

in the simply connected region R if and only if

$$f = \frac{\partial \Omega}{\partial Y}, \qquad g = -\frac{\partial \Omega}{\partial X} \qquad (5.4.2)$$

in R.

Since the equilibrium equations $(5.3.7)_{1,2}$ are of the form

(5.4.1), they imply the existence of scalar fields α and β such that

$$S_{XX} = \frac{\partial \alpha}{\partial Y}, \qquad S_{XY} = -\frac{\partial \alpha}{\partial X} \qquad (5.4.3)$$

$$S_{YY} = \frac{\partial \beta}{\partial X}, \qquad S_{XY} = -\frac{\partial \beta}{\partial Y} \qquad (5.4.4)$$

and so, from $(5.4.3)_2$ and $(5.4.4)_2$, α and β must satisfy

$$\frac{\partial \alpha}{\partial X} = \frac{\partial \beta}{\partial Y} \qquad (5.4.5)$$

Moreover, since equation (5.4.5) is also of the form (5.4.1) this implies the existence of a scalar field ϕ such that

$$\alpha = \frac{\partial \phi}{\partial Y}, \qquad \beta = \frac{\partial \phi}{\partial X} \qquad (5.4.6)$$

Hence,

$$S_{XX} = \frac{\partial^2 \phi}{\partial Y^2}, \quad S_{YY} = \frac{\partial^2 \phi}{\partial X^2}, \quad S_{XY} = -\frac{\partial^2 \phi}{\partial X \, \partial Y} \qquad (5.4.7)$$

and ϕ is known as the *Airy stress function*. Using (5.3.3) we see that

$$\hat{\Theta} = \nabla^2 \phi \qquad (5.4.8)$$

and so $(5.3.7)_3$ becomes

$$\nabla^4 \phi = 0 \qquad (5.4.9)$$

where ∇^4 stands for the double Laplacian operator

$$\nabla^4 = \nabla^2 (\nabla^2) = \frac{\partial^4}{\partial X^4} + 2 \frac{\partial^4}{\partial X^2 \, \partial Y^2} + \frac{\partial^4}{\partial Y^4} \qquad (5.4.10)$$

Equation (5.4.9) is known as the *biharmonic equation* and ϕ is called a *biharmonic function*. Any stress system of the form (5.4.7), where ϕ satisfies (5.4.9), satisfies the system of equations (5.3.7), and conversely all solutions of (5.3.7) must take the form (5.4.7) with ϕ satisfying (5.4.9).

In Sections 5.5 to 5.8 we indicate how this equation can be solved using complex-variable techniques. There are other methods of solving the biharmonic equation, one of which is indicated in the examples at the end of the chapter.

Example 5.4.1

Show that for the case when the body force is conservative, so that $\rho_0\mathbf{b} = -\text{grad } V(X, Y)$, equation (5.4.9) becomes

$$(1-\nu)\,\nabla^4\phi + (1-2\nu)\,\nabla^2 V = 0$$

and that this reduces to (5.4.9) for the case of deformation under a constant gravitational force.

5.5   Complex representation of the Airy stress function

We now show how solutions of the biharmonic equation may be expressed in terms of two analytic functions of a complex variable. This representation is particularly useful in deriving the associated displacement field, and in the solution of boundary-value problems.

We first notice that, in view of (5.4.9), $\nabla^2\phi$ is a plane harmonic function, and so there exists an analytic complex function f such that

$$\nabla^2\phi(X, Y) = \text{Re }\{f(z)\}, \qquad z = X + iY \tag{5.5.1}$$

(see Phillips (1957), p. 16). Now suppose that the region R of the XY-plane occupied by the body, is simply connected, and let z_0 be any point of R. We define

$$\psi(z) = \tfrac{1}{4}\int_{z_0}^{z} f(w)\,\mathrm{d}w \tag{5.5.2}$$

the integral being evaluated along any path from z_0 to z in R. Hence,

$$f = 4\frac{\mathrm{d}\psi}{\mathrm{d}z} \tag{5.5.3}$$

Let p and q be the real and imaginary parts of ψ, so that

$$\psi = p + iq \tag{5.5.4}$$

Since the function ψ must be analytic, p and q satisfy the Cauchy–Riemann equations

$$\frac{\partial p}{\partial X} = \frac{\partial q}{\partial Y}, \qquad \frac{\partial p}{\partial Y} = -\frac{\partial q}{\partial X} \tag{5.5.5}$$

from which it follows that they are harmonic functions, so that

$$\nabla^2\{\phi - pX - qY\} = \nabla^2\phi - 2\left(\frac{\partial p}{\partial X} + \frac{\partial q}{\partial Y}\right) \qquad (5.5.6)$$

But from $(5.5.1)_1$, $(5.5.3)$ and $(5.5.4)$,

$$4\frac{\partial p}{\partial X} = \text{Re}\,\{f\} = \nabla^2\phi \qquad (5.5.7)$$

Hence, from $(5.5.5)_1$ and $(5.5.6)$,

$$\nabla^2\{\phi - pX - qY\} = 0$$

Therefore, there exists an analytic function χ such that

$$\phi - pX - qY = \text{Re}\,\{\chi(z)\}$$

Hence

$$\phi = \text{Re}\,\{\bar{z}\psi(z) + \chi(z)\} \qquad (5.5.8)$$

where a bar over a quantity denotes the complex conjugate: $\bar{z} = X - iY$. Thus, any biharmonic function ϕ may be represented in the form (5.5.8), and conversely, since the real and imaginary parts of any analytic function are harmonic, it is easy to show that any function of the form (5.5.8) is biharmonic.

It is useful to be able to determine the stress components directly from the complex form (5.5.8) for ϕ, and we show now how this can be done in a particularly simple manner.

We see from (5.5.8) that

$$2\phi = \bar{z}\psi(z) + z\,\overline{\psi(z)} + \chi(z) + \overline{\chi(z)} \qquad (5.5.9)$$

Bearing in mind that $z = X + iY$, and that

$$\frac{\partial}{\partial X}\overline{f(X+iY)} = \overline{\frac{\partial f(z)}{\partial z}}, \quad i\frac{\partial}{\partial Y}\overline{f(X+iY)} = \overline{\frac{\partial f(z)}{\partial z}} \qquad (5.5.10)$$

we see that

$$2\frac{\partial\phi}{\partial X} = \psi(z) + \bar{z}\psi'(z) + \overline{\psi(z)} + z\,\overline{\psi'(z)} + \chi'(z) + \overline{\chi'(z)}, \qquad (5.5.11)$$

where $\psi'(z)$ stands for $d\psi/dz$, and

$$2i\frac{\partial\phi}{\partial Y} = \psi(z) - \bar{z}\psi'(z) - \overline{\psi(z)} + z\,\overline{\psi'(z)} - \chi'(z) + \overline{\chi'(z)} \qquad (5.5.12)$$

Hence,

$$\frac{\partial \phi}{\partial X} + i \frac{\partial \phi}{\partial Y} = \psi(z) + z \overline{\psi'(z)} + \overline{\chi'(z)} \qquad (5.5.13)$$

Using (5.4.7),

$$-S_{YY} + iS_{XY} = -\frac{\partial}{\partial X}\left(\frac{\partial \phi}{\partial X} + i \frac{\partial \phi}{\partial Y}\right)$$

$$= -\psi'(z) - \overline{\psi'(z)} - z \overline{\psi''(z)} - \overline{\chi''(z)} \qquad (5.5.14)$$

Now, using (5.3.3), (5.4.8), (5.5.1), and (5.5.3),

$$S_{XX} + S_{YY} = \hat{\Theta} = 2\{\psi'(z) + \overline{\psi'(z)}\} \qquad (5.5.15)$$

and from (5.5.14) and (5.5.15) we can show that

$$\hat{\Phi} \equiv S_{XX} - S_{YY} + 2iS_{XY} = -2\{z \overline{\psi''(z)} + \overline{\chi''(z)}\} \qquad (5.5.16)$$

When the functions ψ and χ are known, (5.5.15) and (5.5.16) provide a particularly simple means of calculating the corresponding stress components.

Example 5.5.1

The cross-section of a cylinder is the region $|X| \leq a$, $|Y| \leq b$. Examine the plane strain–stress system associated with the complex functions

$$\psi(z) = \tfrac{1}{2}Tz, \qquad \chi(z) = 0$$

where T is a real constant.

In this case, using (5.5.15) and (5.5.16) we have

$$S_{XX} + S_{YY} = 2T, \qquad S_{XX} - S_{YY} + 2iS_{XY} = 0$$

Hence

$$S_{XX} = S_{YY} = T, \qquad S_{XY} = 0$$

and the cylinder can be held in this state of uniform stress only by the application of a normal stress T on each of the faces $X = \pm a$, $Y = \pm b$. Note also from (5.3.4) that a normal stress

$$S_{ZZ} = 2\nu T$$

must also act in the Z-direction.

Example 5.5.2

Examine the stress system in the same cylinder corresponding to the complex functions

$$\psi(z) = 0, \quad \chi(z) = \tfrac{1}{2}iSz^2$$

where S is a real constant.

5.6 Determination of the displacement field

We now show how the displacement field may be expressed in terms of the complex functions $\psi(z)$ and $\chi(z)$. Using (5.3.2), (5.3.3), and (5.4.7), we see that the displacement components u and v are related to the Airy stress function ϕ by

$$2\mu \frac{\partial u}{\partial X} = -\frac{\partial^2 \phi}{\partial X^2} + (1 - \nu) \nabla^2 \phi,$$

$$2\mu \frac{\partial v}{\partial Y} = -\frac{\partial^2 \phi}{\partial Y^2} + (1 - \nu) \nabla^2 \phi \qquad (5.6.1)$$

$$\mu \left(\frac{\partial u}{\partial Y} + \frac{\partial v}{\partial X} \right) = -\frac{\partial^2 \phi}{\partial X \partial Y}$$

From (5.5.5) and (5.5.7) we can write $(5.6.1)_{1,2}$ as

$$2\mu \frac{\partial u}{\partial X} = -\frac{\partial^2 \phi}{\partial X^2} + 4(1 - \nu) \frac{\partial p}{\partial X}, \qquad 2\mu \frac{\partial v}{\partial Y} = -\frac{\partial^2 \phi}{\partial Y^2} + 4(1 - \nu) \frac{\partial q}{\partial Y}$$

Integrating, we find

$$2\mu u = -\frac{\partial \phi}{\partial X} + 4(1 - \nu)p + r, \qquad 2\mu v = -\frac{\partial \phi}{\partial Y} + 4(1 - \nu)q + s \qquad (5.6.2)$$

where r depends only on Y, and s depends only on X, and using $(5.5.5)_2$, and $(5.6.1)_3$ we have

$$\frac{dr}{dY} + \frac{ds}{dX} = 0 \qquad (5.6.3)$$

Thus the displacement components (r, s) correspond to zero strain and therefore represent only a rigid-body displacement which does not affect the stresses.

Now using (5.5.4) and (5.5.13) we see that, apart from the

rigid-body displacement,

$$2\mu(u+iv)=(3-4\nu)\psi(z)-z\overline{\psi'(z)}-\overline{\chi'(z)} \qquad (5.6.4)$$

Example 5.6.1

Find the displacement components u and v associated with the complex functions ψ and χ given in Examples 5.5.1 and 5.5.2.

Example 5.6.2

Show that the displacement components corresponding to

$$\psi(z)=iCz+A, \qquad \chi(z)=Bz+D$$

where C is a real constant, and A, B, and D are complex constants, represent a rigid-body displacement [cf. equation (1.7.12)].

5.7   Force on a section of the boundary

Suppose that the arc AB of a curve constitutes part of the boundary of the cross-section. The total force acting on AB has a particularly simple form in terms of the functions ψ and χ, and this is often helpful in the solution of problems. Let $\mathbf{N}$ denote the unit outward-drawn normal to AB, and suppose that $\mathbf{N}$ makes an angle θ with the X-axis. If the arc AB has parametric representation

$$X = X(s), \qquad Y = Y(s)$$

where the arc length s is measured from A to B, then

$$\cos\theta = \frac{dY}{ds}, \qquad \sin\theta = -\frac{dX}{ds} \qquad (5.7.1)$$

The traction $\mathbf{s(N)}$ on the surface has components

$$s_X = S_{XX}\cos\theta + S_{XY}\sin\theta, \qquad s_Y = S_{XY}\cos\theta + S_{YY}\sin\theta \qquad (5.7.2)$$

using (4.3.5). In view of (5.4.7) and (5.7.1) equations (5.7.2) can be written

$$\begin{aligned}
s_X &= \frac{\partial^2\phi}{\partial Y^2}\frac{dY}{ds} + \frac{\partial^2\phi}{\partial X\partial Y}\frac{dX}{ds} = \frac{d}{ds}\left(\frac{\partial\phi}{\partial Y}\right) \\
s_Y &= -\frac{\partial^2\phi}{\partial X^2}\frac{dX}{ds} - \frac{\partial^2\phi}{\partial X\partial Y}\frac{dY}{ds} = -\frac{d}{ds}\left(\frac{\partial\phi}{\partial X}\right)
\end{aligned} \qquad (5.7.3)$$

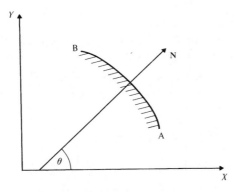

Fig. 5.1

Hence,

$$s_X + is_Y = \frac{d}{ds}\left(\frac{\partial \phi}{\partial Y} - i\frac{\partial \phi}{\partial X}\right) = -i\frac{d}{ds}\left(\frac{\partial \phi}{\partial X} + i\frac{\partial \phi}{\partial Y}\right)$$

Suppose that the total force on the arc AB has components (P, Q), then in complex form

$$P + iQ = \int_A^B (s_X + is_Y)\, ds = -i\int_A^B d\left(\frac{\partial \phi}{\partial X} + i\frac{\partial \phi}{\partial Y}\right)$$
$$= -i\left[\frac{\partial \phi}{\partial X} + i\frac{\partial \phi}{\partial Y}\right]_A^B$$

where the square bracket denotes the difference in value from A to B. Hence, using (5.5.13), we have

$$P + iQ = -i[\psi(z) + z\,\overline{\psi'(z)} + \overline{\chi'(z)}]_A^B \qquad (5.7.4)$$

In particular, if the boundary is free of applied traction then the complex functions must satisfy

$$\psi(z) + z\,\overline{\psi'(z)} + \overline{\chi'(z)} = \text{constant} \qquad (5.7.5)$$

on the boundary. In view of Example 5.6.2, the addition of a constant to $\psi(z)$ corresponds only to a rigid-body displacement. Thus, in a simply connected region the constant on the right-hand side of (5.7.5) may be taken to be zero.

Example 5.7.1

Show that the moment M about the origin of the surface traction acting on the arc AB is given by

$$M = \text{Re}\,[\chi(z) - z\chi'(z) - z\bar{z}\psi'(z)]_A^B \qquad (5.7.6)$$

Now

$$M = \int_A^B (Xs_Y - Ys_X)\,\mathrm{d}s$$

and so, using (5.7.3),

$$M = -\int_A^B \left\{ X\frac{\mathrm{d}}{\mathrm{d}s}\left(\frac{\partial\phi}{\partial X}\right) + Y\frac{\mathrm{d}}{\mathrm{d}s}\left(\frac{\partial\phi}{\partial Y}\right) \right\}\,\mathrm{d}s$$

Integrating by parts, this gives

$$M = -\left[X\frac{\partial\phi}{\partial X} + Y\frac{\partial\phi}{\partial Y} \right]_A^B + \int_A^B \left(\frac{\partial\phi}{\partial X}\frac{\mathrm{d}X}{\mathrm{d}s} + \frac{\partial\phi}{\partial Y}\frac{\mathrm{d}Y}{\mathrm{d}s} \right)\,\mathrm{d}s$$

$$= \left[-X\frac{\partial\phi}{\partial X} - Y\frac{\partial\phi}{\partial Y} + \phi \right]_A^B$$

Using (5.5.13) we find that

$$X\frac{\partial\phi}{\partial X} + Y\frac{\partial\phi}{\partial Y} = \text{Re}\left\{ (X+\mathrm{i}Y)\left(\frac{\partial\phi}{\partial X} - \mathrm{i}\frac{\partial\phi}{\partial Y}\right) \right\}$$

$$= \text{Re}\,\{ z\,\overline{\psi(z)} + z\bar{z}\psi'(z) + z\chi'(z) \}$$

Thus, in view of (5.5.9),

$$M = \text{Re}\,[\{ -z\,\overline{\psi(z)} - z\bar{z}\psi'(z) - z\chi'(z) + z\,\overline{\psi(z)} + \chi(z) \}]_A^B$$

$$= \text{Re}\,[\chi(z) - z\chi'(z) - z\bar{z}\psi'(z)]_A^B$$

Example 5.7.2

Show that on a traction-free boundary the displacement components are given by

$$\mu(u + \mathrm{i}v) = 2(1 - \nu)\psi(z) \qquad (5.7.7)$$

apart from a possible rigid-body displacement.

5.8   Equivalence of function pairs ψ, χ

So far, we have shown that the stress and displacement components in a plane-strain problem may be expressed in terms of the complex functions ψ and χ. We now enquire whether there exists an equivalent pair of functions ψ_1 and χ_1 which corresponds to the same stress and displacement fields (apart from rigid-body displacements).

In view of (5.5.15) and (5.5.16), the pairs of functions ψ, χ and ψ_1, χ_1 give rise to the same stress fields provided

$$\text{Re } \psi'(z) = \text{Re } \psi_1'(z) \tag{5.8.1}$$

$$z\,\overline{\psi''(z)} + \overline{\chi''(z)} = z\,\overline{\psi_1''(z)} + \overline{\chi_1''(z)} \tag{5.8.2}$$

Now it can easily be shown from the Cauchy–Riemann equations that if an analytic function of z is purely imaginary then that function must be a constant. Hence (5.8.1) implies

$$\psi'(z) = \psi_1'(z) + iC \tag{5.8.3}$$

where C is a real constant, and so

$$\psi(z) = \psi_1(z) + iCz + A \tag{5.8.4}$$

where A is a complex constant.

Substituting (5.8.4) in (5.8.2), we find

$$\chi''(z) = \chi_1''(z)$$

and, hence,

$$\chi(z) = \chi_1(z) + Bz + D \tag{5.8.5}$$

where B and D are complex constants.

Using Example 5.6.2, we now see that the displacement fields associated with the two pairs of complex functions differ by, at most, a rigid-body displacement.

5.9 Generalised plane stress

Plane strain is usually discussed in relation to a body which extends indefinitely in the Z-direction, but is bounded by cylindrical surfaces with generators parallel to the Z-axis; moreover, the displacement is independent of Z, and is directed parallel to the XY-plane.

A two-dimensional situation which occurs more commonly in practice is that in which a thin plate is deformed in its own plane. Here we develop an approximate theory for the stretching of thin plates, and show that the governing equations are almost identical with those of plane strain.

Suppose the plate occupies the region

$$(X, Y) \in R, \qquad -h \leqslant Z \leqslant h \qquad (5.9.1)$$

We use the superposed $\sim$ to denote thickness averages. That is, for any function f defined over the region of the plate

$$\tilde{f}(X, Y) = \frac{1}{2h} \int_{-h}^{h} f(X, Y, Z) \, \mathrm{d}Z \qquad (5.9.2)$$

We suppose that the faces of the plate are free of traction, that is,

$$S_{XZ} = S_{YZ} = S_{ZZ} = 0 \qquad \text{on } Z = \pm h \qquad (5.9.3)$$

the plate being deformed by forces in its plane applied to the edges. Then for a sufficiently thin plate it seems reasonable to suppose that the average values of S_{XZ}, S_{YZ}, and S_{ZZ} across the thickness are small compared with S_{XX}, S_{YY}, and S_{XY}, and so we derive our approximate theory on the assumption that

$$\tilde{S}_{XZ} = \tilde{S}_{YZ} = \tilde{S}_{ZZ} = 0 \qquad (5.9.4)$$

Now, assuming that the body force is zero, the equations of equilibrium (4.3.6) integrated across the thickness of the plate give

$$\frac{\partial}{\partial X} \tilde{S}_{XX} + \frac{\partial}{\partial Y} \tilde{S}_{XY} + \frac{1}{2h} [S_{XZ}]_{-h}^{h} = 0$$

$$\frac{\partial}{\partial X} \tilde{S}_{XY} + \frac{\partial}{\partial Y} \tilde{S}_{YY} + \frac{1}{2h} [S_{YZ}]_{-h}^{h} = 0 \qquad (5.9.5)$$

$$\frac{\partial}{\partial X} \tilde{S}_{XZ} + \frac{\partial}{\partial Y} \tilde{S}_{YZ} + \frac{1}{2h} [S_{ZZ}]_{-h}^{h} = 0$$

In view of (5.9.3) and (5.9.4), equation $(5.9.5)_3$ is satisfied identically and $(5.9.5)_{1,2}$ become

$$\frac{\partial}{\partial X} \tilde{S}_{XX} + \frac{\partial}{\partial Y} \tilde{S}_{XY} = 0$$

$$\frac{\partial}{\partial X} \tilde{S}_{XY} + \frac{\partial}{\partial Y} \tilde{S}_{YY} = 0 \qquad (5.9.6)$$

Using Example 4.3.1 we see that the stress–strain relations may be written in the form

$$2\mu E_{KL} = S_{KL} - \nu' S_{MM} \delta_{KL} \qquad (5.9.7)$$

where

$$v' = \lambda/(3\lambda + 2\mu) \qquad (5.9.8)$$

Integrating these through the thickness gives

$$\left. \begin{array}{l} 2\mu \tilde{E}_{XX} = \tilde{S}_{XX} - v'\tilde{\Theta} \\ 2\mu \tilde{E}_{YY} = \tilde{S}_{YY} - v'\tilde{\Theta} \\ 2\mu \tilde{E}_{XY} = \tilde{S}_{XY} \end{array} \right\} \qquad (5.9.9)$$

$$2\mu \tilde{E}_{XZ} = 2\mu \tilde{E}_{YZ} = 0 \qquad (5.9.10)$$

$$(\mu/h)[w]_{-h}^{h} = -v'\tilde{\Theta} \qquad (5.9.11)$$

where $\tilde{\Theta} = \tilde{S}_{XX} + \tilde{S}_{YY}$.

Now equations (5.9.6) and (5.9.9) characterise *generalised plane stress*, and are identical to equations (5.1.3) and (5.3.2), typifying plane strain, except that the stress and strain components in plane strain are here replaced by their average values, and the constant v in plane strain is replaced by v' in generalised plane stress. Having solved a problem in plane strain, (5.3.4) may be used to find the value of S_{ZZ} in terms of $\hat{\Theta}$, whereas in generalised plane stress (5.9.11) relates $\tilde{\Theta}$ to the change in thickness $[w]_{-h}^{h}$ of the plate.

Example 5.9.1

Using $(4.8.3)_3$, show that $v' = v/(1 + v)$ and use $(4.8.4)_3$ to show that $0 < v' < 1/3$.

5.10 Formulation of boundary-value problems

In plane strain the stress and displacement fields are expressible in the forms:

$$\left. \begin{array}{l} S_{XX} + S_{YY} = \hat{\Theta} = 2\{\psi'(z) + \overline{\psi'(z)}\} \\ S_{XX} - S_{YY} + 2iS_{XY} = \hat{\Phi} = -2\{z\overline{\psi''(z)} + \overline{\chi''(z)}\} \end{array} \right\} \qquad (5.10.1)$$

$$2\mu(u + iv) = (3 - 4v)\psi(z) - z\overline{\psi'(z)} - \overline{\chi'(z)} \qquad (5.10.2)$$

where ψ and χ are analytic in the region R, and must be determined subject to certain conditions specified on the boundary. As discussed in Section 4.3, these conditions may involve either the traction or the displacement being specified on the whole of S, or the traction being specified on part of S and the displacement on the remainder. Of course, the traction and displacement on S must be parallel to the XY-plane and independent of Z.

In generalised plane stress, the solution is expressible in the same form (5.10.1) and (5.10.2) except that the stresses and displacement must be replaced by their average values and the constant ν must be replaced by ν'. Boundary-value problems similar to those in plane strain may be considered except that, in place of traction being specified on S, its mean value per unit length of C is given, and in place of the displacement its thickness average is specified.

In the problems discussed later, we normally state the conditions in the context of generalised plane stress, and the symbols for stress and displacement, without the superposed $\sim$, are used to represent average values. Of course, the solution may be easily interpreted in plane strain in the light of the correspondences indicated above.

Some general methods of solution, whereby the functions ψ and χ may be obtained from the boundary conditions, are available; but for these we must refer the reader to other texts (see, for example, Muskhelishvili (1953)). Here we confine our attention to problems which may be solved by inverse methods; that is, by postulating certain forms for ψ and χ and showing that these forms satisfy the required boundary conditions.

Finally, we mention that in our discussion so far we have assumed that R is a bounded, simply connected region. It is, however, possible to deal with other regions as we indicate in some of the following examples. A more general analysis of multiply connected, infinite, and semi-infinite regions will also be found in the text mentioned above.

Example 5.10.1

Using the complex stress transformation laws of Example 8 at the end of Chapter 1, show that the stress components S_{RR}, $S_{R\Theta}$ referred to polar coordinates R, Θ are given by

$$S_{RR} + iS_{R\Theta} = \psi'(z) + \overline{\psi'(z)} - \bar{z}\,\overline{\psi''(z)} - \frac{\bar{z}}{z}\,\overline{\chi''(z)} \qquad (5.10.3)$$

If u_R and u_Θ are the components of the displacement in the $\mathbf{e}_R$

and $\mathbf{e}_\Theta$ directions, show that

$$2\mu(u_R+iu_\Theta)=(3-4\nu')e^{-i\Theta}\psi(z)-R\overline{\psi'(z)}-e^{-i\Theta}\overline{\chi'(z)}$$

$$(5.10.4)$$

Using Example 8 and noting that the coordinate curves at the point (R, Θ) are inclined to the XY-axes at an angle Θ,

$$S_{RR}+S_{\Theta\Theta}=\hat\Theta'=\hat\Theta=2\{\psi'(z)+\overline{\psi'(z)}\} \qquad (5.10.5)$$

$$S_{RR}-S_{\Theta\Theta}+2iS_{R\Theta}=\hat\Phi'=e^{-2i\Theta}\hat\Phi$$

$$=-2\frac{\bar z}{z}\{z\,\overline{\psi''(z)}+\overline{\chi''(z)}\} \qquad (5.10.6)$$

Adding these two equations gives the required result (5.10.3).
   Also,

$$u_R=u\cos\Theta-v\sin\Theta,\qquad u_\Theta=-u\sin\Theta+v\cos\Theta$$

so that using (5.10.2) with ν replaced by ν',

$$2\mu(u_R+iu_\Theta)=2\mu e^{-i\Theta}(u+iv)$$

$$=e^{-i\Theta}\{(3-4\nu')\psi(z)-z\,\overline{\psi'(z)}-\overline{\chi'(z)}\}$$

from which (5.10.4) now follows.

Example 5.10.2

A circular plate with boundary $R=a$ is held in equilibrium by a shear stress $S_{R\Theta}=k\sin 2\Theta$, where k is a constant, applied to the boundary, the normal stress S_{RR} on the boundary being zero. Show that complex functions defined by

$$\psi(z)=Az^3,\qquad \chi(z)=0$$

where A is a real constant, may be used to satisfy the boundary conditions, and evaluate A.

Using (5.10.3),

$$S_{RR}+iS_{R\Theta}=3A(z^2+\bar z^2)-6A\bar z^2$$

$$=3A(z^2-\bar z^2)$$

$$=6Aia^2\sin 2\Theta \qquad \text{on } R=a$$

Hence, the boundary conditions are satisfied provided $A=k/6a^2$.

Example 5.10.3

In the previous example, use the formulae (5.7.4) and (5.7.6) to find the force and couple resultant acting on the segment of the boundary for which $0 \leqslant \Theta \leqslant \pi/2$.

Using (5.7.4),

$$P + iQ = -i[\psi(z) + z\,\overline{\psi'(z)}]_{\Theta=0}^{\Theta=\pi/2}$$
$$= -iA[z^3 + 3z\bar{z}^2]$$
$$= -iAa^3[e^{3i\Theta} + 3e^{-i\Theta}]$$
$$= 4Aa^3(-1+i)$$
$$= \tfrac{2}{3}ak(-1+i)$$

so that the components of the force are given by

$$P = -\tfrac{2}{3}ak, \qquad Q = \tfrac{2}{3}ak$$

Using (5.7.6), the resultant moment is

$$M = \mathrm{Re}\,[-z\bar{z}\psi'(z)]_{\Theta=0}^{\Theta=\pi/2}$$
$$= -\mathrm{Re}\,a^2[3Az^2]$$
$$= -3Aa^4[\cos 2\Theta]$$
$$= 6a^4A$$
$$= a^2k$$

5.11   Multiply connected regions

Suppose the plate occupies a multiply connected region S. Then we know from the foregoing analysis that the stress and the displacement fields in any simply connected subregion R of S may still be represented by a pair of analytic complex functions ψ and χ. Consider, for example, the region S of the annulus $a \leqslant |z| \leqslant b$. Then R may be taken to be the whole region S with a cut introduced along the positive X-axis to make it simply connected. The limiting values of ψ and χ as the positive X-axis is approached from above and below may of course be different. However, the associated stress and displacement components must be single-valued in the whole region S, and in general this requirement imposes additional restrictions on the values of ψ and χ in R. The following examples illustrate this point.

Example 5.11.1

An annular plate occupies the region $a \leqslant |z| \leqslant b$, and the edges $|z| = a$ and $|z| = b$ are subjected to uniform pressures p_1 and p_2, respectively. Show that the solution may be represented by the complex functions

$$\psi(z) = Az, \qquad \chi(z) = B \log z \qquad (5.11.1)$$

where A and B are real constants, and evaluate A and B.

Although χ is multi-valued in the region of the annulus, only the single-valued derivatives of χ enter into the formulae (5.10.1) and (5.10.2) for the stress and displacement components. Moreover, χ is analytic in all simply connected subregions of the annulus.

Using (5.10.3), we see that

$$S_{RR} + iS_{R\Theta} = 2A + \frac{B}{z\bar{z}}$$

The right-hand side is clearly real and so $S_{R\Theta}$ vanishes throughout the region. The boundary conditions require

$$S_{RR} = -p_1 \text{ on } R = a, \qquad S_{RR} = -p_2 \text{ on } R = b$$

Hence,

$$2A + \frac{B}{a^2} = -p_1$$

$$2A + \frac{B}{b^2} = -p_2$$

and so

$$A = \frac{p_2 b^2 - p_1 a^2}{2(a^2 - b^2)}, \qquad B = \frac{a^2 b^2 (p_1 - p_2)}{(a^2 - b^2)} \qquad (5.11.2)$$

Example 5.11.2

The edge $|z| = b$ of the annular plate $a \leqslant |z| \leqslant b$ is held fixed, and the edge $|z| = a$ is displaced as a rigid body an infinitesimal distance d in the X-direction. Show that the boundary conditions may be satisfied by the complex functions

$$\psi(z) = A \log z + Bz^2, \qquad \chi(z) = Cz \log z + Dz + F/z \qquad (5.11.3)$$

where A, B, C, D, and F are real constants to be evaluated. Find the resultant force required to effect this displacement.

Using (5.10.2) (with ν replaced by ν' for generalised plane stress),

$$2\mu(u+iv) = k\psi(z) - z\overline{\psi'(z)} - \overline{\chi'(z)}, \qquad k = 3 - 4\nu'$$

Hence,

$$2\mu(u+iv) = k(A\log z + Bz^2) - z\left(\frac{A}{\bar{z}} + 2B\bar{z}\right) \tag{5.11.4}$$
$$- C - C\log\bar{z} - D + \frac{F}{\bar{z}^2}$$

Since $\log\bar{z} = \log(z\bar{z}) - \log z$, for a single-valued displacement field we must have

$$kA + C = 0 \tag{5.11.5}$$

The formulae (5.10.1) for the stresses involve only the first and second derivatives of ψ and the second derivative of χ, and so the stresses are single-valued without further restriction.

Using (5.11.5), (5.11.4) becomes

$$2\mu(u+iv) = kA\log(z\bar{z}) + kBz^2 - A\frac{z}{\bar{z}} - 2Bz\bar{z} - C - D + \frac{F}{\bar{z}^2} \tag{5.11.6}$$

The boundary conditions are

$$\begin{aligned} u + iv &= d &&\text{on} \quad z\bar{z} = a^2 \\ u + iv &= 0 &&\text{on} \quad z\bar{z} = b^2 \end{aligned} \tag{5.11.7}$$

Hence, (5.11.6) gives

$$kA\log a^2 + kBz^2 - A\frac{z^2}{a^2} - 2Ba^2 - C - D + F\frac{z^2}{a^4} = 2\mu d \tag{5.11.8}$$

$$kA\log b^2 + kBz^2 - A\frac{z^2}{b^2} - 2Bb^2 - C - D + F\frac{z^2}{b^4} = 0$$

on $|z| = a$ and $|z| = b$, respectively. Equations (5.11.8) are satisfied provided

$$kA \log a^2 - 2Ba^2 - C - D = 2\mu d$$
$$kA \log b^2 - 2Bb^2 - C - D = 0$$

$$kB - \frac{A}{a^2} + \frac{F}{a^4} = 0 \qquad (5.11.9)$$

$$kB - \frac{A}{b^2} + \frac{F}{b^4} = 0$$

Equations (5.11.5) and (5.11.9) constitute a set of five simultaneous equations for the five constants. In particular, we find that

$$A = 2\mu dk(a^2 + b^2)\Lambda$$

where

$$\Lambda = \{k^2(a^2 + b^2) \log (a^2/b^2) + 2(b^2 - a^2)\}^{-1}$$

Using (5.7.4), the total force on the inner boundary $|z| = a$ is given by

$$-i[\psi(z) + z \overline{\psi'(z)} + \overline{\chi'(z)}]$$

where [ ] denotes the total change in a complete circuit of the boundary. So the total force is

$$-i2\pi i(A - C) = 2\pi(1 + k)A, \qquad \text{using (5.11.5)}$$
$$= 4\mu\pi dk(1 + k)(a^2 + b^2)\Lambda$$

5.12 Infinite regions

When the linear dimensions of the external boundary of a plate are large compared with some internal boundary, we often idealise the problem by considering the plate to be of infinite extent. The functions ψ and χ must still be analytic in any simply connected, bounded subregion, and the only additional consideration is their behaviour as $|z| \to \infty$. Again we refer the reader to Muskhelishvili (1953), Chapter 17, for a general discussion of the restrictions on ψ and χ arising from the conditions at infinity. Here we simply illustrate the procedure with some examples.

Example 5.12.1

An infinite plate contains a circular hole $|z| = a$ which is free of applied traction. When a uniform pressure p is applied at infinity, show that the state of stress in the plate may be described by complex functions

$$\psi(z) = Az, \qquad \chi(z) = B \log z \qquad (5.12.1)$$

where A and B are real constants to be evaluated. Find the peripheral stress on the edge of the hole. Show also that the deformed hole is circular and find its radius.

This example is the same as Example 5.11.1 in the limit as $b \to \infty$ and with $p_1 = 0$, $p_2 = p$. Hence, from (5.11.2),

$$A = -\tfrac{1}{2}p, \qquad B = a^2 p.$$

Using (5.10.3) and (5.10.5), we find

$$S_{RR} = -p\left(1 - \frac{a^2}{R^2}\right), \quad S_{R\Theta} = 0, \quad S_{\Theta\Theta} = -p\left(1 + \frac{a^2}{R^2}\right) \quad (5.12.2)$$

The peripheral stress on $|z| = a$ is therefore $S_{\Theta\Theta} = -2p$.

Using (5.7.7), we see that the displacement components on $|z| = a$ are given by

$$\mu(u + iv) = 2(1 - v)(-\tfrac{1}{2}pz) = -(1 - v)pz$$

We note that they are single-valued and

$$u = -\bar{p}X, \quad v = -\bar{p}Y, \quad \bar{p} = (1 - v)p/\mu \qquad (5.12.3)$$

Thus

$$x = X + u = X(1 - \bar{p}), \qquad y = Y + v = Y(1 - \bar{p})$$

and so the circle $X^2 + Y^2 = a^2$ is transformed into

$$x^2 + y^2 = a^2(1 - \bar{p})^2$$

that is, another circle but with radius $a(1 - \bar{p})$.

Example 5.12.2

An infinite plate contains a circular hole $|z| = a$ which is free of applied traction. A uniaxial tension $S_{XX} = T$, $S_{XY} = 0$, $S_{YY} = 0$ is applied at infinity. Show that complex functions of the form

$$\psi(z) = Az + \frac{B}{z}, \qquad \chi(z) = Cz^2 + D \log z + \frac{F}{z^2} \qquad (5.12.4)$$

where A, B, C, D, and F are real constants, may be used to satisfy the boundary conditions of the problem. Find the peripheral stress and the deformed configuration of the hole.

Using (5.10.1) we see that, as $|z| \to \infty$,

$$S_{XX} + S_{YY} = 4A, \qquad S_{XX} - S_{YY} + 2iS_{XY} = -4C$$

so that

$$S_{XX} = 2(A - C), \qquad S_{YY} = 2(A + C), \qquad S_{XY} = 0$$

Hence, the conditions at infinity may be satisfied by taking

$$A = \tfrac{1}{4}T, \qquad C = -\tfrac{1}{4}T \tag{5.12.5}$$

Using (5.7.5), we see that the boundary conditions on the hole are satisfied if

$$Az + \frac{B}{z} + z\left(A - \frac{B}{\bar{z}^2}\right) + 2C\bar{z} + \frac{D}{\bar{z}} - 2\frac{F}{\bar{z}^3} = 0$$

when $z\bar{z} = a^2$. Thus

$$\left(2A + \frac{D}{a^2}\right)z + (B + 2Ca^2)\frac{1}{z} - \left(\frac{B}{a^4} + \frac{2F}{a^6}\right)z^3 = 0$$

when $z = ae^{i\Theta}$ for all Θ. This condition implies

$$2A + \frac{D}{a^2} = 0, \qquad B + 2Ca^2 = 0, \qquad Ba^2 + 2F = 0 \tag{5.12.6}$$

Equations (5.12.5) and (5.12.6) imply

$$D = -\tfrac{1}{2}a^2 T, \qquad B = \tfrac{1}{2}a^2 T, \qquad F = -\tfrac{1}{4}a^4 T \tag{5.12.7}$$

and note from (5.10.2) that the displacement is single-valued.

To find the peripheral stress we note that on the hole $S_{RR} = 0$ and so

$$\begin{aligned}
S_{\Theta\Theta} = S_{RR} + S_{\Theta\Theta} &= 2\{\psi'(z) + \overline{\psi'(z)}\} \\
&= 2\left\{2A - B\left(\frac{1}{z^2} + \frac{1}{\bar{z}^2}\right)\right\} \\
&= T(1 - 2\cos 2\Theta)
\end{aligned}$$

using (5.12.5) and (5.12.6)$_2$. This means that the peripheral stress varies between $-T$ on the X-axis and $3T$ on the Y-axis (see Fig. 5.2).

Using (5.7.7), the displacement components on the hole are

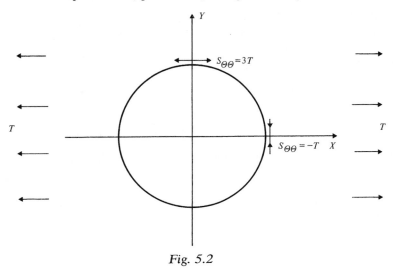

Fig. 5.2

given by

$$\mu(u+iv) = 2(1-\nu)\psi(z) = (1-\nu)T\left(\tfrac{1}{2}z + \frac{a^2}{z}\right)$$

so that

$$u+iv = (1-\nu)(T/\mu)\{\tfrac{1}{2}(X+iY)+X-iY\}$$

and

$$x = X+u = \{1+3(1-\nu)T/2\mu\}X$$
$$y = Y+v = \{1-(1-\nu)T/2\mu\}Y$$

Hence, the circle $X^2 + Y^2 = a^2$ deforms into the ellipse

$$\frac{x^2}{a^2\{1+3(1-\nu)T/2\mu\}^2} + \frac{y^2}{a^2\{1-(1-\nu)T/2\mu\}^2} = 1$$

which has semi-major and minor axes in the x- and y-directions with lengths

$$a\{1+3(1-\nu)T/2\mu\}, \quad a\{1-(1-\nu)T/2\mu\}$$

respectively (see Fig. 5.3).

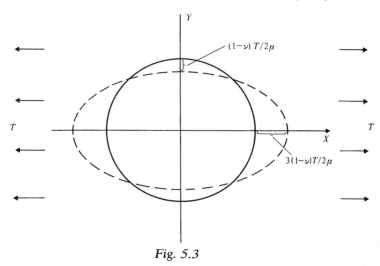

Fig. 5.3

5.13   Isolated point force

Here we consider certain complex functions with singularities at the origin, and show how they may be used to represent the effect of an isolated point force. We remind the reader of our remarks in Section 4.11 concerning the relation between the singular stress field and the limiting solution obtained by allowing the region of application of a distributed force to shrink to zero.

Suppose that the origin is an interior point of the region R, and consider the complex functions

$$\psi(z) = A \log z, \qquad \chi(z) = Bz \log z \qquad (5.13.1)$$

where A and B are complex constants. Using (5.6.4), we see that, for the displacement to be single-valued, we require

$$(3 - 4\nu)A + \bar{B} = 0 \qquad (5.13.2)$$

Using (5.7.4), we see that the resultant force on any boundary enclosing the origin is $-(P + iQ)$, where

$$-(P + iQ) = -i2\pi i(A - \bar{B}) = 8\pi A(1 - \nu) \qquad (5.13.3)$$

This force resultant is clearly independent of the size and shape of the boundary, and so we regard the complex functions

$$\psi(z) = -\frac{(P + iQ)}{8\pi(1 - \nu)} \log z, \qquad \chi(z) = \frac{(3 - 4\nu)(P - iQ)}{8\pi(1 - \nu)} z \log z \qquad (5.13.4)$$

as being associated with a complex isolated force $P + iQ$ at the origin.

Example 5.13.1

Show that the resultant couple, arising from the complex functions (5.13.4), acting on any boundary enclosing the origin is zero.

Example 5.13.2

Show that the complex functions

$$\psi(z) = 0, \qquad \chi(z) = (G/2\pi)\mathrm{i} \log z \qquad (5.13.5)$$

represent an isolated couple of magnitude G acting at the origin.

Example 5.13.3

The surface of a circular cylinder $|z| \leq a$ is held fixed and an isolated couple of magnitude G per unit length acts along the axis. Show that the stress system may be described by complex functions

$$\psi(z) = Az, \qquad \chi(z) = (G/2\pi)\mathrm{i} \log z$$

provided $A = \mathrm{i}G/8\pi(1-\nu)a^2$.

Examples 5

1. An elastic circular cylinder occupies the region $R \leq a$. The surface $R = a$ is subjected to a displacement $w = \varepsilon \cos \Theta$ in the Z-direction, where ε is a constant, and body force is absent. Find the displacement in the whole cylinder and the surface traction on $R = a$ required to produce the displacement.

2. The cross-section of an elastic plate is the region $0 \leq X \leq a$, $|Y| \leq b$. The major faces $Y = \pm b$ are free of applied traction. On the edge $X = a$ the normal stress is zero but the shear stress gives rise to a resultant force of magnitude F per unit length in the Z-direction and acts in the positive Y-direction. Tractions are applied to the edge $X = 0$ to balance the force F. Assuming that a state of plane strain exists in planes $Z = \text{constant}$ and that body force is absent, find suitable real values of the constants A, B,

and C, such that the function

$$\phi = \tfrac{1}{6}AY^3 + \tfrac{1}{6}BXY^3 + CXY$$

is biharmonic and satisfies the boundary conditions on $X = a$ and $Y = \pm b$. Show that the resultant bending moment per unit length in the Z-direction on $X = c$ is $F(c - a)$.

3. An infinite elastic plate contains a circular hole $|z| = a$ which is free of applied traction. A pure shear $S_{XX} = 0$, $S_{YY} = 0$, $S_{XY} = S$ is applied at infinity. Show that the complex functions

$$\psi(z) = i\,\frac{A}{z}, \qquad \chi(z) = iBz^2 + i\,\frac{C}{z^2}$$

where A, B, and C are real constants, may be used to satisfy the boundary conditions. Find the peripheral stress on the edge of the hole, and show that the deformed boundary has equation

$$(1 + \alpha^2)x^2 - 4\alpha xy + (1 + \alpha^2)y^2 = (1 - \alpha^2)^2 a^2$$

where $\alpha = 2(1 - \nu')S/\mu$, representing an ellipse.

4. An elastic circular cylinder of radius b contains a coaxial, rigid, cylindrical inclusion of radius a which adheres at all points on its boundary to the elastic material. The inclusion is given an infinitesimal rotation through an angle α about its axis, while the outer cylindrical surface is held fixed. Show that the complex functions

$$\psi(z) = Az, \qquad \chi(z) = B \log z$$

where A and B are complex constants, may be used to satisfy the boundary conditions.

Find the moment of the couple per unit length which must be applied to the inner cylinder to effect this rotation.

5. An infinite elastic plate contains a circular hole $z\bar{z} = a^2$. The stress vector acting on the edge of the plate has a constant value all the way round, and the resultant force on the edge has components (P, Q). Show that on $z\bar{z} = a^2$

$$S_{RR} + iS_{R\Theta} = -\frac{1}{2\pi a}(P + iQ)e^{-i\Theta}$$

Show also that the complex functions

$$\psi(z) = A \log z, \qquad \chi(z) = Bz \log z + \frac{C}{z}$$

may be used to describe the state of stress, provided

$$A = -\frac{(P+iQ)}{2\pi(1+k)}, \quad B = \frac{k(P-iQ)}{2\pi(1+k)}, \quad C = -\frac{a^2(P+iQ)}{2\pi(1+k)} \qquad (k = 3-4\nu')$$

Show that the stresses at any point z, in the limit as $a \to 0$ keeping P and Q fixed, are the same as those arising from the complex functions (5.13.4) representing the isolated-force solution, suitably modified for generalised plane stress.

6. Elastic material occupies the half-space $Y \geqslant 0$. Show that the stress components, in plane polar coordinates, arising from the complex functions

$$\psi(z) = -\frac{iQ}{2\pi} \log z, \quad \chi(z) = -\frac{iQ}{2\pi} z \log z$$

are

$$S_{RR} = -\frac{2Q}{\pi R} \sin \Theta, \qquad S_{R\Theta} = S_{\Theta\Theta} = 0, \quad R \neq 0$$

Deduce that these complex functions are associated with an isolated point force of magnitude Q at the origin directed along the Y-axis, the boundary $Y = 0$ being free of traction.

Extension, torsion, and bending

6.1 The deformation of long cylinders

In this chapter we discuss again the deformation of cylinders of elastic material in equilibrium, but here the deformation is assumed to arise from tractions applied to the end sections. We suppose that the generators of the surface S of the cylinders are parallel to the Z-axis, and that the end sections E_0 and E_l lie in the planes $Z = 0$ and $Z = l$, respectively. We consider the surface S to be free of applied traction; the deformation is produced entirely by tractions applied to the ends E_0 and E_l.

The exact solution of this three-dimensional problem is one of considerable analytical complexity. However, in many practical situations, the cylinder may be regarded as long compared with the linear dimensions of a cross-section and, except for small regions near E_0 and E_l, the deformation is determined almost entirely by the force and couple resultants associated with the tractions on E_0 and E_l, and is not dependent on the manner in which the traction is distributed. We therefore consider some particularly simple deformations for which the resultant forces and couples are easily calculated. Away from the end regions, these deformations may be taken as the solutions to any practical problem with the same end resultants. The precise interpretation to be given to our use of the terms 'long cylinder' and 'small region' involves a more detailed discussion of Saint-Venant's principle than is possible within the scope of this volume. The interested reader is referred to Gurtin (1972), Sections 54 to 56.

In the absence of body force, the systems of force and couple resultants acting on E_0 and E_l must clearly be self-equilibrating, and we now construct separate displacement fields for which the end-section resultants consist of:

(i) forces of equal magnitude F, but opposite sense, parallel to the Z-axis (extension);

(ii) couples of equal magnitude M, but opposite sense, with vector moment parallel to the Z-axis (torsion);

(iii) couples of equal magnitude M, but opposite sense, with vector moment parallel to the Y-axis (bending by couples).

6.2   Extension

As in the previous chapter, we suppose that a cross-section of the cylinder occupies a region R of the XY-plane and that the area of R is A. Consider the case of uniaxial tension discussed in Section 4.7 (ii) with

$$S_{ZZ} = F/A. \tag{6.2.1}$$

That is, we take displacement components

$$u = -(\nu F/AE)X, \quad v = -(\nu F/AE)Y, \quad w = (F/AE)Z \tag{6.2.2}$$

Then, as shown in Section 4.7, the only non-vanishing stress is S_{ZZ}, and so the conditions for a traction-free surface S are identically satisfied. Moreover, the distribution of traction S_{ZZ} over E_l is equivalent to a single force at $(0, 0, l)$ together with a couple. The force is clearly in the Z-direction and of magnitude

$$\int_R S_{ZZ}\, dS = F, \tag{6.2.3}$$

and the moment of the couple has components N_X, N_Y given by

$$N_X = \int_R YS_{ZZ}\, dS = \frac{F}{A} \int_R Y\, dS$$

$$N_Y = -\int_R XS_{ZZ}\, dS = -\frac{F}{A} \int_R X\, dS$$

We remind the reader here that the *centroid* of a plane region R of area A in the XY-plane is the point with coordinates $(\bar{X}, \bar{Y})$ where

$$A\bar{X} = \int_R X\, dS, \qquad A\bar{Y} = \int_R Y\, dS$$

If, therefore, we take the Z-axis to be the line of centroids of the cross-sections, then $N_X = N_Y = 0$, and the traction on E_l is just equivalent to a force F along the Z-axis. Likewise, the traction on E_0 can be shown to be equivalent to a force $-F$ along the Z-axis.

6.3 Torsion of a circular cylinder

We begin our discussion of the torsion problem by considering the displacement components for the simple torsion field of Section 1.6 for small values of the angle of twist τ. From (1.6.1), (1.6.4), and (1.7.1) we see that

$$u_1 = (c-1)X_1 - sX_2, \quad u_2 = sX_1 + (c-1)X_2, \quad u_3 = 0 \quad (6.3.1)$$

where $s = \sin \tau X_3$ and $c = \cos \tau X_3$. Thus

$$u_1 = -\tau X_3 X_2 + O(\tau^2), \quad u_2 = \tau X_3 X_1 + O(\tau^2), \quad u_3 = 0 \quad (6.3.2)$$

as $\tau \to 0$. Now, writing (u, v, w) in place of (u_1, u_2, u_3) and (X, Y, Z) in place of (X_1, X_2, X_3), we consider in our infinitesimal theory the displacement components of order τ in (6.3.2):

$$u = -\tau ZY, \quad v = \tau ZX, \quad w = 0 \qquad (6.3.3)$$

The only non-vanishing stresses associated with this displacement field are

$$S_{ZX} = -\mu\tau Y, \qquad S_{ZY} = \mu\tau X \qquad (6.3.4)$$

and these clearly satisfy the equations of equilibrium (4.3.6) in the absence of body force. The condition that S is free of applied traction implies that $\mathbf{s(N)} = \mathbf{0}$ on S, where $\mathbf{N}$ is the unit normal to S. That is, using (4.3.5),

$$lS_{ZX} + mS_{ZY} = 0 \qquad (6.3.5)$$

where $\mathbf{N} = (l, m, 0)$. In the case of a circular cylinder,

$$l = X/R, \qquad m = Y/R \qquad (6.3.6)$$

where

$$R = \sqrt{(X^2 + Y^2)}$$

and so, using (6.3.4) and (6.3.6), we see that (6.3.5) is identically satisfied.

The only non-vanishing component of the moment of the

couple acting on E_l is in the Z-direction and is of magnitude

$$M = \int_R (X S_{ZY} - Y S_{ZX}) \, \mathrm{d}S$$

$$= \mu\tau \int_R (X^2 + Y^2) \, \mathrm{d}S$$

$$= \mu\tau \int_0^{2\pi} \int_0^a R^3 \, \mathrm{d}R \, \mathrm{d}\Theta$$

$$= \tfrac{1}{2}\mu\tau\pi a^4$$

A couple of equal magnitude but opposite sense is easily shown to act on E_0.

Example 6.3.1

Show that the resultant force on the end sections is zero. [This is in contrast with the situation in finite elasticity where normal forces on the end sections are necessary to maintain simple torsion (see Section 3.10).]

Example 6.3.2

Show that the boundary condition (6.3.5) is satisfied by the stress components (6.3.4) *only* when the boundary of the cross-section is a circle.

6.4   Torsion of cylinders of arbitrary cross-section

We consider now a cylinder, the surface of which meets the XY-plane in an arbitrary smooth curve C bounding a region R. Denoting the outward unit normal to C by (l, m), the condition that the surface of the cylinder is free of applied traction again gives

$$l S_{ZX} + m S_{ZY} = 0 \qquad \text{on } C \qquad (6.4.1)$$

We saw in the previous section that the stress components associated with the displacement field (6.3.3) satisfy this boundary condition identically when C is a circle centred at the origin. However, to solve the torsion problem when the boundary of the cross-section is an arbitrary smooth curve C, we require a more general displacement field, and we consider the effect of allowing the cross-sections of the cylinder to 'warp' as they rotate. In other words, we suppose that during rotation each point of a cross-section may be displaced in the Z-direction, the magnitude of the displacement varying over the section, but being the same function of (X, Y) for each cross-section. This means that we consider displacement components of the form

$$u = -\tau ZY, \quad v = \tau ZX, \quad w = \tau\phi(X, Y) \tag{6.4.2}$$

where ϕ is known as the *warping function*. The only non-vanishing stresses are now found to be

$$S_{ZX} = \mu\tau\left(\frac{\partial\phi}{\partial X} - Y\right), \quad S_{ZY} = \mu\tau\left(\frac{\partial\phi}{\partial Y} + X\right) \tag{6.4.3}$$

Substituting (6.4.3) into the equilibrium equations (4.3.6) in the absence of body force, we find that for $L = 1, 2$ the equations are satisfied identically. The equation in which $L = 3$ gives

$$\nabla^2\phi = \frac{\partial^2\phi}{\partial X^2} + \frac{\partial^2\phi}{\partial Y^2} = 0 \tag{6.4.4}$$

The boundary condition (6.4.1) becomes

$$l\left(\frac{\partial\phi}{\partial X} - Y\right) + m\left(\frac{\partial\phi}{\partial Y} + X\right) = 0 \qquad \text{on } C \tag{6.4.5}$$

Since ϕ is a plane harmonic function, there exists a conjugate plane-harmonic function ψ related to ϕ by the Cauchy–Riemann equations

$$\frac{\partial\phi}{\partial X} = \frac{\partial\psi}{\partial Y}, \qquad \frac{\partial\phi}{\partial Y} = -\frac{\partial\psi}{\partial X} \tag{6.4.6}$$

To show how such a function ψ may be found, choose any point $A(X_0, Y_0)$ of R and take the value of ψ at A to be some arbitrary value ψ_A. Then the value of ψ at a neighbouring point $B(X_0 + dX, Y_0 + dY)$ is given by

$$\psi_B = \psi_A + \frac{\partial\psi}{\partial X}dX + \frac{\partial\psi}{\partial Y}dY$$

$$= \psi_A - \frac{\partial\phi}{\partial Y}dX + \frac{\partial\phi}{\partial X}dY \tag{6.4.7}$$

Hence, a value of ψ at any other point $P(X, Y)$ of R is given by

$$\psi = \psi_A + \int_A^P \left(-\frac{\partial \phi}{\partial Y} \, dX + \frac{\partial \phi}{\partial X} \, dY \right) \qquad (6.4.8)$$

where the line integral is taken over any curve in R joining A and P. Using Stokes's theorem we see that, for any closed curve c bounding a region s contained in R,

$$\oint_c \left(-\frac{\partial \phi}{\partial Y} \, dX + \frac{\partial \phi}{\partial X} \, dY \right) = \int_s \nabla^2 \phi \, dS = 0 \qquad (6.4.9)$$

and so the function ψ defined by (6.4.8) is single-valued provided R is simply connected.

However, if R is not simply connected then the function ψ as defined in (6.4.8) may be multi-valued. Suppose, for example, that a simple closed curve C_1 completely encloses another simple closed curve C_2, and R is taken to be the region between C_1 and C_2. In other words, we are considering the torsion of a hollow cylinder. Then it is possible to construct closed curves c in R (see Fig. 6.1), enclosing C_2 so that the region s bounded by such a curve c is not entirely in R; equation (6.4.9) is then not necessarily valid since the integrand $\nabla^2 \phi$ is undefined over part of s. In this case, the definition (6.4.8) of ψ may depend on the path of integration from A to P; different values of ψ may be obtained according to the number of times the path encloses C_2. This difficulty may be removed by introducing a notional 'cut' along a line joining C_1 and C_2; R may then be regarded as simply connected and ψ single-valued. A similar device may be used for other kinds of multiply connected regions.

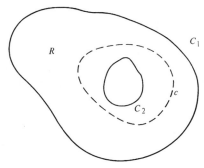

Fig. 6.1

Returning now to the boundary condition (6.4.5), we see that, using (6.4.6), we may write it in the form

$$l\frac{\partial\psi}{\partial Y} - m\frac{\partial\psi}{\partial X} - lY + mX = 0 \qquad \text{on } C \qquad (6.4.10)$$

Now let $\boldsymbol{\sigma}$ denote the unit tangent to C, so that $\boldsymbol{\sigma} = (-m, l)$. Equation (6.4.10) can be written

$$(\boldsymbol{\sigma}\cdot\nabla)\{\psi - \tfrac{1}{2}(X^2 + Y^2)\} = 0 \qquad \text{on } C \qquad (6.4.11)$$

Equation (6.4.11) implies that $\psi - \tfrac{1}{2}(X^2 + Y^2)$ must be constant on C. Hence, the torsion problem for a cylinder of arbitrary cross-section reduces to that of finding a plane harmonic function ψ which takes the value $\tfrac{1}{2}(X^2 + Y^2) + \text{constant}$ on C, which is the Dirichlet problem of potential theory.

We recall from (6.4.8) that ψ is defined to within an arbitrary constant only. If R is simply connected, the constant may sometimes conveniently be chosen so that $\psi - \tfrac{1}{2}(X^2 + Y^2)$ vanishes on C. If R is multiply connected, $\psi - \tfrac{1}{2}(X^2 + Y^2)$ can be made to vanish only on one disjoint part of the boundary; on other parts it will take other distinct constant values.

6.5   The Prandtl stress function and the lines of shearing stress

The function

$$\chi = \psi - \tfrac{1}{2}(X^2 + Y^2) \qquad (6.5.1)$$

is known as the *Prandtl stress function*, and the torsion problem may be formulated entirely in terms of χ. Since ψ is a harmonic function, χ must satisfy the equation

$$\nabla^2\chi = -2 \qquad (6.5.2)$$

and in view of (6.4.11) the boundary conditions are satisfied when $\chi = \text{constant}$ on C. (When R is multiply connected, we emphasise again that χ may take different constant values on disjoint parts of the boundary.)

Using (6.4.3), (6.4.6), and (6.5.1), we see that the stress components are now given by the simple relations

$$S_{ZX} = \mu\tau\frac{\partial\chi}{\partial Y}, \qquad S_{ZY} = -\mu\tau\frac{\partial\chi}{\partial X} \qquad (6.5.3)$$

Consider the family of curves $\chi = $ constant. The normal
at any point of such a curve is in the direction of
grad $\chi = (\partial\chi/\partial X, \partial\chi/\partial Y)$; but note that the stress vector acting
at that point on a cross-section has components
$(S_{ZX}, S_{ZY}) = \mu\tau(\partial\chi/\partial Y, -\partial\chi/\partial X)$. The stress vector is therefore
perpendicular to grad χ. Thus the stress vectors acting on a cross-
section are tangential to the curves $\chi = $ constant. These curves
are known as *lines of shearing stress*.

Clearly, for any function χ satisfying (6.5.2) a curve
$\chi = $ constant may be regarded as the boundary of a possible cross-
section for the solution of an associated torsion problem. We may
also take, for example, any two such curves, one completely enclosing the other, as external and internal boundaries of the cross-
section of a hollow cylinder (see Example 6.7.2).

6.6   Maximum shearing stress

Theorem.   The magnitude of the stress vector acting on a cross-
section takes its maximum value on the boundary of the cross-
section.

Proof.   The stress vector acting on a cross-section has no component in the Z-direction since $S_{ZZ} = 0$; and it is therefore entirely
a shear stress, acting parallel to the XY-plane, with magnitude
$(S_{ZX}^2 + S_{ZY}^2)^{\frac{1}{2}}$. Let P be any interior point of R and take the X-
axis to be parallel to the direction of the stress vector at P. Then
the square of the magnitude of this stress vector is S_{ZX}^2, evaluated
at P, and we denote this by $(S_{ZX}^2)_P$.

Now

$$S_{ZX} = \mu\tau\left(\frac{\partial\phi}{\partial X} - Y\right) \tag{6.6.1}$$

and so in view of (6.4.4) we see that

$$\nabla^2 S_{ZX} = 0 \tag{6.6.2}$$

From a theorem in potential theory, S_{ZX} therefore cannot have a
maximum or minimum value at the interior point P of R; and so
a point Q, distinct from P, must exist such that

$$(S_{ZX}^2)_Q \geqslant (S_{ZX}^2)_P \tag{6.6.3}$$

Hence, the square of the magnitude of the total shear stress at Q satisfies the inequality

$$(S_{ZX}^2 + S_{ZY}^2)_Q \geqslant (S_{ZX}^2)_Q \geqslant (S_{ZX}^2)_P \qquad (6.6.4)$$

This means that at no interior point P of R can the magnitude of the total shear stress be a maximum. The maximum must therefore occur on the boundary C.

6.7 Force and couple resultants on a cross-section

Let the force resultant on a cross-section with outward normal in the Z-direction have components (P, Q) in the X- and Y-directions. (Since $S_{ZZ} = 0$ there is clearly no Z-component.) Then

$$P = \int_R S_{ZX}\, dS = \mu\tau \int_R \frac{\partial\chi}{\partial Y}\, dS$$

$$= -\mu\tau \int_R \text{curl}\,(\chi e_X) \cdot e_Z\, dS$$

$$= -\mu\tau \oint_C \chi e_X \cdot d\mathbf{X} \qquad (6.7.1)$$

using Stokes's theorem. Since χ is constant on C and e_X is a constant unit vector, the line integral (6.7.1) must vanish. Hence $P = 0$, and we can show similarly that $Q = 0$. The couple acting on a cross-section has moment M about the Z-axis, where

$$M = \int_R (XS_{ZY} - YS_{ZX})\, dS$$

$$= -\mu\tau \int_R \left(X\frac{\partial\chi}{\partial X} + Y\frac{\partial\chi}{\partial Y} \right) dS \qquad (6.7.2)$$

$$= \mu\tau \int_R \left\{ 2\chi - \frac{\partial}{\partial X}(X\chi) - \frac{\partial}{\partial Y}(Y\chi) \right\} dS$$

$$= 2\mu\tau \int_R \chi\, dS - \mu\tau \oint_C \chi(lX + mY)\, dS \qquad (6.7.3)$$

using the divergence theorem. As discussed in Section 6.4, *when R is simply connected* we may take the constant value of χ on C

to be zero. In this case, (6.7.3) reduces to

$$M = 2\mu\tau \int_R \chi \, \mathrm{d}S \qquad (6.7.4)$$

When R is not simply connected (see Example 6.7.2), the expression (6.7.2) is often easier to evaluate than (6.7.3).

Example 6.7.1

Find the function χ for the cylinder of circular cross-section $X^2 + Y^2 \le a^2$ with $\chi = 0$ on the boundary. Verify using (6.7.4) that the twisting moment is $\frac{1}{2}\mu\tau\pi a^4$.

Example 6.7.2

Show that the twisting moment for a hollow cylinder $b^2 \le X^2 + Y^2 \le a^2$ is $\frac{1}{2}\mu\tau\pi(a^4 - b^4)$.

These examples illustrate that for a hollow cylinder the torsional rigidity M/τ is less than that for the corresponding solid cylinder.

6.8   Torsion of a cylinder with elliptical cross-section

We have shown that the solution of a torsion problem reduces to that of finding a function χ which satisfies

$$\nabla^2 \chi = -2 \qquad (6.8.1)$$

within R, and is constant on the boundary C. Suppose now that C is the ellipse with equation

$$\frac{X^2}{a^2} + \frac{Y^2}{b^2} = 1 \qquad (6.8.2)$$

and consider the function

$$\chi = k\left(\frac{X^2}{a^2} + \frac{Y^2}{b^2} - 1\right) \qquad (6.8.3)$$

where k is a constant. This form of χ clearly satisfies the boundary condition on C since $\chi = 0$ on C. Moreover,

$$\nabla^2 \chi = 2k\left(\frac{1}{a^2} + \frac{1}{b^2}\right)$$

and so (6.8.1) is satisfied if

$$k = -a^2 b^2 / (a^2 + b^2) \qquad (6.8.4)$$

Using (6.7.4) we see that the moment of the twisting couple is

$$M = \frac{2\mu\tau a^2 b^2}{a^2 + b^2} \int_R \left(1 - \frac{X^2}{a^2} - \frac{Y^2}{b^2} \right) dS \qquad (6.8.5)$$

where R is the region $x^2/a^2 + y^2/b^2 \leq 1$. The integral can be evaluated by changing to the variables r and θ where $X = ar \cos\theta$, $Y = br \sin\theta$ for which the Jacobian is abr. We then obtain

$$M = \frac{2\mu\tau a^3 b^3}{a^2 + b^2} \int_0^{2\pi} \int_0^1 (1 - r^2) r \, dr \, d\theta$$

$$= \pi\mu\tau a^3 b^3 / (a^2 + b^2) \qquad (6.8.6)$$

Example 6.8.1

Show that the total shear σ on C is given by

$$\sigma^2 = 4k^2 \mu^2 \tau^2 \left\{ \frac{1}{b^2} - \frac{X^2}{a^2} \left(\frac{1}{b^2} - \frac{1}{a^2} \right) \right\}$$

Deduce that the maximum shear occurs at the points $(0, \pm b)$ if $b < a$.

Example 6.8.2

Find the warping function for the elliptic cylinder, and sketch the contour lines on a cross-section.

Using (6.5.1), (6.8.3), and (6.8.4), we find

$$\psi = \frac{(a^2 - b^2)}{2(a^2 + b^2)} (X^2 - Y^2) + \frac{a^2 b^2}{a^2 + b^2}$$

Hence, using (6.4.6),

$$\frac{\partial \phi}{\partial X} = \frac{\partial \psi}{\partial Y} = -\frac{(a^2 - b^2)}{a^2 + b^2} Y$$

$$\frac{\partial \phi}{\partial Y} = -\frac{\partial \psi}{\partial X} = -\frac{(a^2 - b^2)}{a^2 + b^2} X$$

from which we may deduce that, apart from an arbitrary additive constant,

$$\phi = -\frac{(a^2 - b^2)}{a^2 + b^2} XY$$

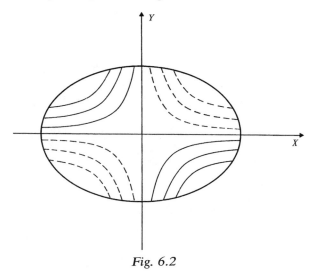

Fig. 6.2

The contours ϕ = constant are therefore hyperbolae with the X-and Y-axes as asymptotes. The displacement in the first and third quadrants is in the negative Z-direction, and in the second and fourth quadrants in the positive Z-direction (see Fig. 6.2).

6.9 Torsion of a cylinder of equilateral-triangular cross-section

Consider the equilateral triangular cross-section shown in Fig. 6.3. The edges have equations

$$X - a = 0, \quad X - Y\sqrt{3} + 2a = 0, \quad X + Y\sqrt{3} + 2a = 0 \quad (6.9.1)$$

Therefore the function

$$\chi = k(X - a)(X - Y\sqrt{3} + 2a)(X + Y\sqrt{3} + 2a) \quad (6.9.2)$$

where k is a constant, clearly vanishes at each point of the boundary C. Moreover, this value of χ may be written

$$\chi = k(X - a)\{(X + 2a)^2 - 3Y^2\} \quad (6.9.3)$$

After deriving the second partial derivatives, we find

$$\nabla^2 \chi = 12ka$$

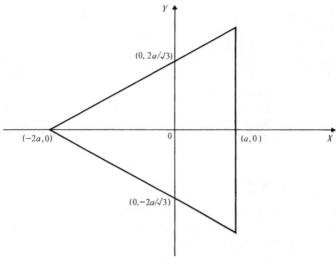

Fig. 6.3

Thus $\nabla^2\chi = -2$ if we take

$$k = \frac{-1}{6a} \qquad (6.9.4)$$

Using (6.5.3), we find

$$S_{ZX} = \frac{\mu\tau Y}{a}(X-a), \quad S_{ZY} = \frac{\mu\tau}{2a}\{X(X+2a) - Y^2\}$$

Now on $X = a$, $S_{ZX} = 0$ and the total shear stress is

$$S_{ZY} = \frac{\mu\tau}{2a}(3a^2 - Y^2)$$

By the symmetry of the equilateral triangle the maximum shear stress therefore occurs at the mid-points of the edges (that is, when $Y = 0$, on $X = a$) and has magnitude $\frac{3}{2}\mu\tau a$.

Example 6.9.1

Use the formula (6.7.4) to show that $M = (9\sqrt{3}/5)\mu\tau a^4$.

Example 6.9.2

Show that the warping function for the triangular cross-section is $(6a)^{-1}(3X^2Y - Y^3)$ and sketch the contour lines.

We have considered two cross-sections for which there exist simple solutions for χ. There are other shapes for which solutions can be found by a similar approach, some of which appear in the examples at the end of the chapter. For some cross-sections, for example, squares and rectangles, the form for χ is in terms of an infinite series. Complex-variable techniques can also be employed in the solution of torsion problems. But we do not consider this approach here. The interested reader is referred to Muskhelishvili (1953), Chapter 22.

6.10   Bending by terminal couples

We first remind the reader that the moments and product of inertia about the X- and Y-axes of a uniform lamina of unit density, occupying the region R, are

$$I_X = \int_R Y^2 \, dS, \quad I_Y = \int_R X^2 \, dS, \quad H_{XY} = \int_R XY \, dS \quad (6.10.1)$$

and that it is possible to choose X- and Y-axes so that $H_{XY} = 0$. These axes are then known as *principal axes of inertia*, and I_X, I_Y are the *principal moments of inertia*. (see Yeh and Abrams (1960), p. 263).

Suppose that, as in the case of extension, we take the Z-axis along the line of centroids of the cross-sections, and that we take X- and Y-axes to be principal axes of inertia of the cross-section. We now want to find the displacement in the beam when it is *bent* by the application of couples on the end sections. To begin with, we suppose that the couple acting on E_l has moment M about the Y-axis (Fig. 6.4). That is,

$$-\int_R XS_{ZZ} \, dS = M \tag{6.10.2}$$

the moment about the X-axis being zero, that is,

$$\int_R YS_{ZZ} \, dS = 0 \tag{6.10.3}$$

As before, we take the major surface S of the cylinder to be traction-free, so that

$$\mathbf{s}(\mathbf{N}) = \mathbf{0} \tag{6.10.4}$$

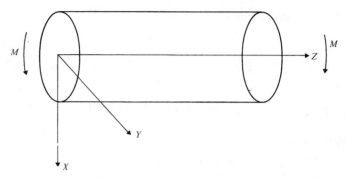

Fig. 6.4

on S, where $\mathbf{N} = (l, m, 0)$ is the unit normal to S. We also require the resultant force on the end sections to be zero; that is,

$$\int_R \mathbf{s}(\mathbf{e}_Z) \, dS = \mathbf{0} \qquad (6.10.5)$$

The simplest stress field which appears capable of satisfying conditions (6.10.2) to (6.10.5) is

$$S_{ZZ} = kX, \; S_{XX} = S_{YY} = S_{YZ} = S_{ZX} = S_{XY} = 0 \qquad (6.10.6)$$

where k is a constant. This stress field clearly satisfies the equilibrium equations (4.3.6) in the absence of body force, and using (4.7.22) the associated strain components are

$$E_{XX} = E_{YY} = -(\nu k/E)X, \quad E_{ZZ} = (k/E)X \qquad (6.10.7)$$

for which the strain compatibility equations (4.5.15) are satisfied identically. An associated displacement field does therefore exist, and we have a possible solution of the equations of elasticity. With regard to the boundary conditions, we note that (6.10.4) requires

$$lS_{XX} + mS_{YX} = 0, \quad lS_{XY} + mS_{YY} = 0, \quad lS_{XZ} + mS_{YZ} = 0 \quad (6.10.8)$$

and in view of (6.10.6), these relations are satisfied identically. Conditions (6.10.3) and (6.10.5) are satisfied because of our choice of axes described at the beginning of this section. Condition (6.10.2) requires

$$-\int_R kX^2 \, dS = M \qquad (6.10.9)$$

and, using $(6.10.1)_2$, $(6.10.9)$ gives

$$k = -M/I_Y \qquad (6.10.10)$$

It can easily be verified directly that the displacement field

$$u = \tfrac{1}{2}m(Z^2 + \nu X^2 - \nu Y^2), \quad v = m\nu XY, \quad w = -mXZ \qquad (6.10.11)$$

where $m = M/EI_Y$, gives rise to the stress distribution $(6.10.6)$.

Example 6.10.1

Using $(6.10.7)$ deduce by integration that, apart from an arbitrary rigid-body displacement, the displacement field corresponding to the stress distribution $(6.10.6)$ is given by $(6.10.11)$.

From definition $(4.3.4)$ of the strain components it follows from $(6.10.7)$ that the displacement components satisfy

$$\frac{\partial u}{\partial X} = \nu mX, \quad \frac{\partial v}{\partial Y} = \nu mX, \quad \frac{\partial w}{\partial Z} = -mX \qquad (6.10.12)$$

$$\frac{\partial v}{\partial X} + \frac{\partial u}{\partial Y} = 0, \quad \frac{\partial w}{\partial Y} + \frac{\partial v}{\partial Z} = 0, \quad \frac{\partial u}{\partial Z} + \frac{\partial w}{\partial X} = 0 \qquad (6.10.13)$$

By integration of $(6.10.12)$ we see that u, v, w must take the forms

$$u = \tfrac{1}{2}\nu mX^2 + u_0(Y, Z), \quad v = \nu mXY + v_0(Z, X),$$
$$w = -mZX + w_0(X, Y) \qquad (6.10.14)$$

Substituting these expressions in $(6.10.13)$ we find

$$\nu mY + \frac{\partial v_0}{\partial X}(Z, X) + \frac{\partial u_0}{\partial Y}(Y, Z) = 0 \qquad (6.10.15)$$

$$\frac{\partial w_0}{\partial Y}(X, Y) + \frac{\partial v_0}{\partial Z}(Z, X) = 0 \qquad (6.10.16)$$

$$\frac{\partial u_0}{\partial Z}(Y, Z) - mZ + \frac{\partial w_0}{\partial X}(X, Y) = 0 \qquad (6.10.17)$$

From $(6.10.15)$ we conclude that v_0 is linear in X; from $(6.10.16)$ that w_0 is linear in Y and v_0 linear in Z; and from $(6.10.17)$ that w_0 is linear in X. We therefore write

$$v_0 = \gamma X - \alpha Z + b, \qquad w_0 = \alpha Y - \beta X + c \qquad (6.10.18)$$

where α, β, γ, b, and c are constants, and the same constant α appears in v_0 and w_0 in order to satisfy (6.10.16). Using (6.10.18) in (6.10.15) and (6.10.16), we find

$$\frac{\partial u_0}{\partial Y}(Y, Z) = -\nu m Y - \gamma$$

$$\frac{\partial u_0}{\partial Z}(Y, Z) = mZ + \beta$$

(6.10.19)

from which we conclude that u_0 must have the form

$$u_0 = \tfrac{1}{2}m(Z^2 - \nu Y^2) + \beta Z - \gamma Y + a$$ (6.10.20)

where a is a constant. Substituting (6.10.18) and (6.10.20) into (6.10.14), we see that, apart from the terms involving the constants α, β, γ and a, b, c, the displacement components are those given in (6.10.11). From Section 1.7 we see that the terms involving α, β, γ correspond to an infinitesimal rigid-body rotation, and those involving a, b, c arise from a rigid-body translation.

6.11   The Euler–Bernoulli law

From (6.10.11) we see that on the line of material, or fibre, lying along the Z-axis ($X = 0$, $Y = 0$) the only non-zero displacement component is

$$u = \tfrac{1}{2}mZ^2$$ (6.11.1)

Since for this fibre $u = x$ and $w = 0$ so that $z = Z$ the equation of the fibre after the deformation is

$$x = \tfrac{1}{2}mz^2$$ (6.11.2)

This means that this fibre is in the xz-plane, and assumes the form of a parabola with vertex at the origin and axis lying along the x-axis.

Now the radius of curvature R of this parabola is given by

$$\frac{1}{R} = \frac{\mathrm{d}^2 x}{\mathrm{d}z^2}\left\{1 + \left(\frac{\mathrm{d}x}{\mathrm{d}z}\right)^2\right\}^{-\frac{3}{2}} = m(1 + m^2 z^2)^{-\frac{3}{2}}$$ (6.11.3)

Hence, to the first order in mz, that is, near to the origin, the curvature of the bar is

$$1/R = m = M/EI_Y.$$ (6.11.4)

This means that, for a given bar, the curvature is proportional to the bending moment M. The quantity EI_Y is called the *flexural rigidity* of the beam. Equation (6.11.4) is generally known as the *Euler–Bernoulli law.*

Of course, a similar solution may be constructed to give a bending moment about the X-axis, and by addition to the above solution, a bending moment about any axis in the XY-plane, through the origin, can be produced.

In the above solution, the beam has a constant bending moment M on every cross-section. This is in contrast with the situation considered in Example 2 at the end of Chapter 5 in which the bending moment varied along the length of the beam.

Examples 6

1. Using plane polar coordinates (R, Θ), show that the function

$$\chi = -\tfrac{1}{2}(R^2 - b^2)\left(1 - \frac{2a \cos \Theta}{R}\right)$$

satisfies

$$\nabla^2 \chi = -2$$

where $R \neq 0$, and a, b are constants $(2a > b)$.

Sketch the curves along which $\chi = 0$ and show that χ may be used to describe the torsion stress system in an elastic circular shaft of radius a along which a circular cylindrical groove of radius b has been cut.

Find the stresses S_{XZ}, S_{YZ} on the boundary, and show that the total shear on the shaft surface is

$$\mu\tau\left(a - \frac{b^2}{4a} \sec^2 \Theta\right)$$

while the total shear on the groove surface is

$$\mu\tau(2a \cos \Theta - b)$$

Find the point at which the total shear is greatest and show that, at this point, its value is $(\mu\tau/4a)(4a^2 - b^2)$.

2. Show that, for a given twist τ and given cross-sectional area, the torsion moment M for a cylinder of elliptical cross-section is greatest when the ellipse is a circle.

3. Show that the stress function

$$\chi = -\frac{a^2b^2}{a^2+b^2}\left(\frac{X^2}{a^2}+\frac{Y^2}{b^2}-1\right)$$

may be used to describe the torsion stress system in a hollow cylinder the cross-section of which is bounded by the similar ellipses

$$\frac{X^2}{a^2}+\frac{Y^2}{b^2}=1, \quad \frac{X^2}{a^2}+\frac{Y^2}{b^2}=k^2, \quad 0<k<1.$$

Find the torsion moment M for this cross-section.

4. When f and g are twice continuously differentiable scalar fields, Green's theorem may be written in the form

$$\int_{V_0} (f\,\nabla^2 g + \text{grad}\,f \cdot \text{grad}\,g)\,dV_0 = \int_{S_0} f\mathbf{N} \cdot \text{grad}\,g\,dS_0$$

where the region V_0 is bounded by the closed surface S_0 with outward unit normal $\mathbf{N}$. Use this theorem to show that

$$\int_R \left(X\frac{\partial\psi}{\partial X} + Y\frac{\partial\psi}{\partial Y}\right)dS = \tfrac{1}{2}\int_C (X^2+Y^2)\mathbf{N} \cdot \nabla\psi\,ds$$

where R is a region in the XY-plane bounded by the curve C, $\mathbf{N}$ is the outward unit normal to C, and ψ is a plane harmonic function.

Use the theorem again, together with the boundary condition (6.4.11), to show that the torsion moment for the cross-section R may be written in the form

$$M = \mu\tau\int_R (X^2+Y^2)\,dS - \mu\tau\int_R \left\{\left(\frac{\partial\psi}{\partial X}\right)^2 + \left(\frac{\partial\psi}{\partial Y}\right)^2\right\}dS,$$

and deduce that $M \leq \mu\tau I_0$, where I_0 is the moment of inertia of the cross-section R about a perpendicular axis through O. Show that $M = \mu\tau I_0$ only when R is a circle or annulus.

Elastic waves

So far we have considered problems in which the bodies are in equilibrium so that $\mathbf{u} = \mathbf{u}(\mathbf{X})$. An important dynamic application of the infinitesimal theory of elasticity in which $\mathbf{u} = \mathbf{u}(\mathbf{X}, t)$ is to the study of small-amplitude vibrations. The particular application of this subject upon which we concentrate here is the propagation of seismic waves. These waves contain information concerning the internal composition of the Earth since the energy released at the source, or 'focus', of an earthquake radiates in all directions, travelling through the whole of the interior of the Earth including the core. By interpreting the information received at seismological stations on the Earth's surface, the seismologist, in effect, produces an internal picture of the Earth's structure.

Oldham (1900) identified on seismograms three main types of waves. They were (i) primary waves which are compression and expansion waves like sound waves in a gas; (ii) secondary waves arising from vibrations which are perpendicular to the direction of travel, similar in this respect to light waves; and (iii) surface waves which appear in the upper 30 km or so of the Earth's crust. Since the first two types of wave occur within the interior of the Earth, they are called *body waves*. To analyse these waves mathematically, seismologists regard the Earth as a homogeneous, isotropic, elastic solid to which the infinitesimal theory of elasticity is applicable. However, in reality the Earth consists of layers of different materials, and so we include a discussion of some of the effects of these layers on wave propagation. Body force is largely insignificant, and we therefore neglect it. We also neglect the Earth's curvature and take all surfaces to be horizontal planes; this is sufficiently accurate for earthquakes near the surface, but for deep earthquakes curvature should be considered.

7.1   One-dimensional wave equation. Notion of a plane wave

Consider the one-dimensional wave equation

$$V^2 \frac{\partial^2 \phi}{\partial X^2} = \frac{\partial^2 \phi}{\partial t^2} \tag{7.1.1}$$

where $\phi = \phi(X, t)$ represents some physical quantity. In our earlier notation the right-hand side of (7.1.1) should be $D^2\phi/Dt^2$. However, to maintain the familiar form of the wave equation, throughout this chapter we use the notation $\partial/\partial t$ to denote the partial derivative with respect to time even though quantities are still expressed in terms of material coordinates.

Introducing the new variables

$$\xi = t - X/V, \qquad \eta = t + X/V$$

and using the chain rule of partial differentiation, we find

$$\frac{\partial \phi}{\partial X} = \frac{\partial \phi}{\partial \xi}\frac{\partial \xi}{\partial X} + \frac{\partial \phi}{\partial \eta}\frac{\partial \eta}{\partial X} = -\frac{1}{V}\frac{\partial \phi}{\partial \xi} + \frac{1}{V}\frac{\partial \phi}{\partial \eta}$$

$$\frac{\partial \phi}{\partial t} = \frac{\partial \phi}{\partial \xi}\frac{\partial \xi}{\partial t} + \frac{\partial \phi}{\partial \eta}\frac{\partial \eta}{\partial t} = \frac{\partial \phi}{\partial \xi} + \frac{\partial \phi}{\partial \eta}$$

and

$$\frac{\partial^2 \phi}{\partial X^2} = \frac{\partial}{\partial \xi}\left(\frac{\partial \phi}{\partial X}\right)\frac{\partial \xi}{\partial X} + \frac{\partial}{\partial \eta}\left(\frac{\partial \phi}{\partial X}\right)\frac{\partial \eta}{\partial X} = \frac{1}{V^2}\left\{\frac{\partial^2 \phi}{\partial \xi^2} - 2\frac{\partial^2 \phi}{\partial \xi\,\partial \eta} + \frac{\partial^2 \phi}{\partial \eta^2}\right\}$$

$$\frac{\partial^2 \phi}{\partial t^2} = \frac{\partial}{\partial \xi}\left(\frac{\partial \phi}{\partial t}\right)\frac{\partial \xi}{\partial t} + \frac{\partial}{\partial \eta}\left(\frac{\partial \phi}{\partial t}\right)\frac{\partial \eta}{\partial t} = \frac{\partial^2 \phi}{\partial \xi^2} + 2\frac{\partial^2 \phi}{\partial \xi\,\partial \eta} + \frac{\partial^2 \phi}{\partial \eta^2}$$

Thus (7.1.1) becomes

$$\frac{\partial^2 \phi}{\partial \xi\,\partial \eta} = 0$$

where ϕ is now regarded as a function of ξ and η. Hence,

$$\frac{\partial \phi}{\partial \eta} = G(\eta)$$

where $G(\eta)$ is an arbitrary function of η. A further integration gives

$$\phi = f(\xi) + g(\eta)$$

where $f(\xi)$ is an arbitrary function of ξ and $g(\eta) = \int G(\eta)\,d\eta$. In terms of the original variables X and t, we have

$$\phi = f(t - X/V) + g(t + X/V) \qquad (7.1.2)$$

which is known as *D'Alembert's solution* of the wave equation (7.1.1). The forms of f and g are determined by the initial and boundary conditions.

In order to interpret the solution (7.1.2) physically, consider first the expression

$$\phi = f(t - X/V)$$

At time $t = t_0$, $\phi = f(t_0 - X/V)$. At time $t = t_0 + \delta$, $\phi = f(t_0 + \delta - X/V)$, that is $\phi = f(t_0 - X'/V)$ where $X' = X - V\delta$. Thus ϕ depends on X at time $t = t_0$ in the same manner as ϕ depends on X' at time $t = t_0 + \delta$. This means that there is no change in the waveform, but since $X = X' + V\delta$ the waveform has moved a distance $V\delta$ in the positive X-direction. This is shown diagrammatically in Fig. 7.1.

Since at any instant f depends only on X, it has the same value at all points in any plane normal to the X-axis. Thus the disturbance represented by $f(t - X/V)$ is a *progressive plane wave*. Similar reasoning shows that the term $g(t + X/V)$ represents a plane wave moving in the negative X-direction with speed of propagation V.

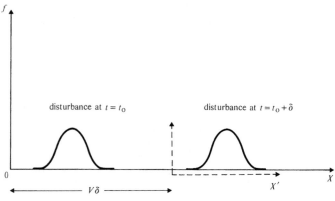

disturbance at $t = t_0$ disturbance at $t = t_0 + \delta$

$V\delta$

X'

X

Fig. 7.1

Example 7.1.1

Show that, in the absence of body force, the displacement field

$$u = u(X, t), \qquad v = w = 0 \qquad (7.1.3)$$

satisfies the equations of motion (4.4.4) provided that u is a solution of the one-dimensional wave equation

$$v_p^2 \frac{\partial^2 u}{\partial X^2} = \frac{\partial^2 u}{\partial t^2} \qquad (7.1.4)$$

where

$$v_p^2 = (\lambda + 2\mu)/\rho_0 \qquad (7.1.5)$$

Show further that for such a displacement field the only non-vanishing components of stress are S_{XX}, S_{YY}, and S_{ZZ} given by

$$S_{XX} = (\lambda + 2\mu) \, \partial u/\partial X, \qquad S_{YY} = S_{ZZ} = \lambda \, \partial u/\partial X \qquad (7.1.6)$$

Example 7.1.2

Elastic material occupies the half-space $X \geq 0$. At time $t = 0$ the material is at rest in its undeformed configuration. For times $t \geq 0$ the surface $X = 0$ is subject to a uniform time-dependent pressure $p(t)$. Assuming that the displacement field is given by (7.1.3), use the results derived in Example 7.1.1 to calculate the displacement and stresses within the material for all times $t \geq 0$.

We have to find the displacement $u(X, t)$ satisfying (7.1.4) and subject to the *boundary conditions*

$$S_{XX} = -p(t), \quad S_{XY} = S_{XZ} = 0 \quad \text{on } X = 0 \text{ for } t \geq 0 \qquad (7.1.7)$$

and the *initial conditions*

$$u = 0, \quad \partial u/\partial t = 0 \quad \text{for } X \geq 0 \text{ at } t = 0 \qquad (7.1.8)$$

Taking the general solution of (7.1.4) in the form (7.1.2), we have in this case

$$u = f(t - X/v_p) + g(t + X/v_p) \qquad (7.1.9)$$

and using $(7.1.6)_1$,

$$S_{XX} = -\rho_0 v_p \{ f'(t - X/v_p) - g'(t + X/v_p) \} \qquad (7.1.10)$$

where $f'(s)$ and $g'(s)$ denote df/ds and dg/ds, respectively. The boundary conditions $(7.1.7)_{2,3}$ are satisfied identically, and $(7.1.7)_1$

gives

$$\rho_0 v_p \{f'(t) - g'(t)\} = p(t), \qquad t \geq 0 \qquad (7.1.11)$$

The initial conditions (7.1.8) give

$$f(-X/v_p) + g(X/v_p) = 0, \qquad X \geq 0 \qquad (7.1.12)$$

$$f'(-X/v_p) + g'(X/v_p) = 0, \qquad X \geq 0 \qquad (7.1.13)$$

Now (7.1.13) implies

$$-f(-X/v_p) + g(X/v_p) = \text{constant} = 2k, \quad \text{say, for } X \geq 0$$

Using (7.1.12) we have

$$-f(-X/v_p) = g(X/v_p) = k, \qquad X \geq 0 \qquad (7.1.14)$$

and so, in particular,

$$-f(0) = g(0) = k \qquad (7.1.15)$$

Integrating (7.1.11), and using (7.1.15), we have

$$f(t) - g(t) = (1/\rho_0 v_p) \int_0^t p(s) \, ds - 2k, \qquad t \geq 0$$

and using (7.1.14) we find

$$f(t) = (1/\rho_0 v_p) \int_0^t p(s) \, ds - k, \qquad t \geq 0 \qquad (7.1.16)$$

In view of (7.1.14) and (7.1.16), the displacement (7.1.9) becomes

$$u = \begin{cases} (1/\rho_0 v_p) \displaystyle\int_0^{t-X/v_p} p(s) \, ds, & t - X/v_p \geq 0 \\ 0, & t - X/v_p < 0 \end{cases} \qquad (7.1.17)$$

and the stress components are

$$S_{XX} = \begin{cases} -p(t - X/v_p), & t - X/v_p \geq 0 \\ 0, & t - X/v_p < 0 \end{cases}$$

$$S_{YY} = S_{ZZ} = \begin{cases} -(\lambda/\rho_0 v_p^2) p(t - X/v_p), & t - X/v_p \geq 0 \\ 0, & t - X/v_p < 0 \end{cases}$$

At any given time t we see that the surface $X = v_p t$ which is called the *wavefront* separates the half-space into a region $X > v_p t$, in which there is no displacement and the material is unstressed, and a region $X < v_p t$ where the displacement is proportional to the area under the pressure curve from 0 to the point $t - X/v_p$. In this region the normal stress S_{xx}, which is compressive, is the pressure applied to the boundary at the previous time $t - X/v_p$. Hence, each pressure signal propagates into the medium with speed v_p. The velocity of the particle at X is

$$\frac{\partial u}{\partial t} = \begin{cases} (1/\rho_0 v_p) p(t - X/v_p), & t - X/v_p \geqslant 0 \\ 0, & t - X/v_p < 0 \end{cases}$$

and is in the same direction as the wave velocity.

So far we have been considering the special case when the disturbance propagates in the X-direction. Let us now consider the three-dimensional wave equation

$$V^2 \nabla^2 \phi = \frac{\partial^2 \phi}{\partial t^2} \tag{7.1.18}$$

and the case of a plane wave propagating in the direction of a unit vector $\mathbf{N}$, called the *propagation vector*. To find the appropriate generalisation of the first term of (7.1.2), let OX' be an axis in the direction of $\mathbf{N}$, then a progressive wave advancing in the X'-direction is represented by

$$\phi = f(t - X'/V) \tag{7.1.19}$$

but $X' = \mathbf{X} \cdot \mathbf{N}$, so that (7.1.19) may be written

$$\phi = f(t - \mathbf{X} \cdot \mathbf{N}/V) \tag{7.1.20}$$

Similarly,

$$\phi = g(t + \mathbf{X} \cdot \mathbf{N}/V) \tag{7.1.21}$$

represents a plane wave propagating with speed V in the $-\mathbf{N}$ direction. The earlier discussion corresponds to the case when $\mathbf{N} = (1, 0, 0)$. Other solutions of (7.1.18) exist which do not correspond to plane waves.

Example 7.1.3

Show by direct substitution that (7.1.20) and (7.1.21) satisfy (7.1.18).

7.2   Plane harmonic waves

When the functions f and g in D'Alembert's solution (7.1.2) are sines or cosines the wave is called *plane harmonic*. These waves

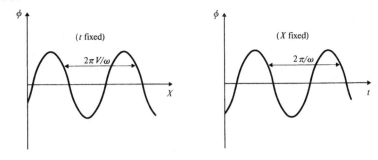

Fig. 7.2

are of fundamental importance in the linear theories of wave
motion since they constitute the basic elements in terms of which
more elaborate solutions may be expressed by the methods of Fourier
analysis.

Consider the plane harmonic wave represented by

$$\phi = A \cos\{\omega(t - X/V) - \alpha\} \qquad (7.2.1)$$

where A is a constant (see Fig. 7.2). We call ω the *angular
frequency*, $\omega/2\pi$ the *frequency*, and α the *phase* of the wave. The
maximum value A of ϕ is called the *wave amplitude*. Now ϕ is
periodic in both X and t. The period in X is $2\pi V/\omega$ and this
is known as the *wavelength*. The period in t is $2\pi/\omega$ and this is
simply referred to as the *period* of the wave; $\omega/2\pi V$ is the *wave
number*.

Since the right-hand side of (7.2.1) can be written as

$$\text{Re}\,[B\exp\{i\omega(t-X/V)\}], \qquad B = Ae^{-i\alpha} \qquad (7.2.2)$$

where Re denotes the real part, for mathematical convenience we
frequently write

$$\phi = B\exp\left\{i\omega\left(t - \frac{X}{V}\right)\right\} \qquad (7.2.3)$$

with the understanding that the real part of the right-hand side is
to be taken when physical interpretation is being made. The appropriate expression for a plane harmonic wave propagating in
the direction of the unit vector $\mathbf{N}$ is

$$\phi = B\exp\left\{i\omega\left(t - \frac{\mathbf{X}\cdot\mathbf{N}}{V}\right)\right\} \qquad (7.2.4)$$

Example 7.2.1

Verify that (7.2.4) satisfies (7.1.18).

7.3   Elastic body waves

In Example 7.1.2 we saw that prescribed initial and boundary conditions gave rise to a plane wave in a half-space travelling in the X-direction with speed v_p. One characteristic of this wave is that the displacement is in the direction of propagation. Other types of wave are possible, and in order to gain further insight into the nature of small-amplitude disturbances in an elastic solid, we consider next an infinite medium in which the displacement depends only on X and t. When $\mathbf{u} = \mathbf{u}(X, t)$ the equations of motion (4.4.3) reduce to

$$\rho_0 \frac{\partial^2 u}{\partial t^2} = (\lambda + 2\mu) \frac{\partial^2 u}{\partial X^2}, \quad \rho_0 \frac{\partial^2 v}{\partial t^2} = \mu \frac{\partial^2 v}{\partial X^2}, \quad \rho_0 \frac{\partial^2 w}{\partial t^2} = \mu \frac{\partial^2 w}{\partial X^2}. \quad (7.3.1)$$

Using (7.1.2) we can write the general solution of these equations in the form

$$u = f_1(t - X/v_p) + g_1(t + X/v_p)$$
$$v = f_2(t - X/v_s) + g_2(t + X/v_s) \quad (7.3.2)$$
$$w = f_3(t - X/v_s) + g_3(t + X/v_s)$$

where

$$v_p = \left(\frac{\lambda + 2\mu}{\rho_0}\right)^{\frac{1}{2}} = \left\{\frac{E(1-\nu)}{\rho_0(1+\nu)(1-2\nu)}\right\}^{\frac{1}{2}}, \quad v_s = \left(\frac{\mu}{\rho_0}\right)^{\frac{1}{2}} = \left\{\frac{E}{2\rho_0(1+\nu)}\right\}^{\frac{1}{2}}$$
$$(7.3.3)$$

Specimen values of v_p and v_s are given in Table 7.1.

Table 7.1

	v_p (m/s)	v_s (m/s)
Aluminium	6.32×10^3	3.07×10^3
Copper	4.36×10^3	2.13×10^3
Iron	5.80×10^3	3.14×10^3

From (7.3.3) we see that

$$v_p = \left(\frac{2(1-\nu)}{1-2\nu}\right)^{\frac{1}{2}} v_s \qquad (7.3.4)$$

and using the inequality $(4.8.4)_3$ we can show that

$$v_p > \sqrt{2}\, v_s \qquad (7.3.5)$$

For many materials, and in particular most rocks of the Earth's crust, λ is approximately equal to μ (so that $\nu = \frac{1}{4}$ approximately). This is known as *Poisson's relation*, and such materials as *Poisson materials*. For these materials $v_p = \sqrt{3}\, v_s$.

In order to analyse the disturbance represented by (7.3.2) we first consider three special cases.

(i) When $u = f_1(t - X/v_p)$, $v = w = 0$, the disturbance consists of a progressive plane wave travelling with speed v_p in the positive X-direction. Since the displacement is everywhere parallel to the direction of propagation, the wave is said to be *longitudinal*. Further $e = \text{div } \mathbf{u} \neq 0$, so that there are volume changes, and the wave is called a *dilatational wave*, or in seismology, a *P-wave*. We also note that for this wave, curl $\mathbf{u} = \mathbf{0}$.

(ii) When $u = 0$, $v = f_2(t - X/v_s)$, $w = 0$, the disturbance consists of a progressive plane wave travelling with speed v_s in the positive X-direction. Since the displacement is everywhere perpendicular to the direction of propagation being confined to the planes $X =$ constant, the wave is said to be *transverse*. For this disturbance $e = 0$, curl $\mathbf{u} \neq \mathbf{0}$, and the wave is called a *rotational wave, shear wave*, or in seismology, an *S-wave*. The displacement and the direction of propagation are confined to planes $Z =$ constant and the wave is sometimes said to be *polarised* in these planes.

(iii) When $u = 0$, $v = 0$, $w = f_3(t - X/v_s)$, we have an S-wave polarised in the planes $Y =$ constant.

P and S stand for 'primary' and 'secondary'. Since $v_p > v_s$, the primary waves travel faster than the secondary waves. When an S-wave is polarised so that all the particles move horizontally, it is referred to as an SH-wave. When the particles move vertically, the wave is called an SV-wave.

Combining the above remarks, we see from (7.3.2) that a general plane disturbance consists of six independent progressive plane waves, namely a P-wave, an SH-wave, and an SV-wave travelling in the positive X-direction, and a representative of each wave travelling in the negative X-direction.

7.4   Potential function representation

In Chapter 4 we saw how potential functions can be used to solve static boundary-value problems. A potential representation of the displacement field which is useful in wave-propagation problems is

$$\mathbf{u} = \text{grad } \phi + \text{curl } \boldsymbol{\psi}, \qquad \text{div } \boldsymbol{\psi} = 0 \qquad (7.4.1)$$

where $\phi(\mathbf{X}, t)$ and $\boldsymbol{\psi}(\mathbf{X}, t)$ are scalar and vector fields respectively, known as the *Lamé potentials*. Using (7.4.1), equations (4.4.3) become

$$\text{grad}\left(v_p^2 \nabla^2 \phi - \frac{\partial^2 \phi}{\partial t^2}\right) + \text{curl}\left(v_s^2 \nabla^2 \boldsymbol{\psi} - \frac{\partial^2 \boldsymbol{\psi}}{\partial t^2}\right) = \mathbf{0}$$

so that, provided ϕ and $\boldsymbol{\psi}$ are such that

$$v_p^2 \nabla^2 \phi = \frac{\partial^2 \phi}{\partial t^2}, \qquad v_s^2 \nabla^2 \boldsymbol{\psi} = \frac{\partial^2 \boldsymbol{\psi}}{\partial t^2} \qquad (7.4.2)$$

the equations of motion are satisfied. Since, from (7.4.1)

$$\text{div } \mathbf{u} = \nabla^2 \phi, \qquad \text{curl } \mathbf{u} = \text{curl curl } \boldsymbol{\psi} = -\nabla^2 \boldsymbol{\psi} \qquad (7.4.3)$$

we see that ϕ is associated with a dilatational wave and $\boldsymbol{\psi}$ with a rotational wave. Although every sufficiently smooth vector field allows a decomposition in the form (7.4.1) (see Morse and Feshbach (1953)), it is not obvious that every solution of the field equations (4.4.3) is of this form where the scalar and vector potentials ϕ and $\boldsymbol{\psi}$ satisfy (7.4.2). However, this can be established (see Sternberg (1960) for a historical discussion and proof). Hence, instead of using (4.4.3) we can use (7.4.2), which enable the two types of wave to be analysed separately.

When $\phi = \phi(X, Z, t)$, $\boldsymbol{\psi} = \boldsymbol{\psi}(X, Z, t)$, and writing $\psi = \psi_2$, we have expressions for the first and third components of displacement in the form

$$u = \frac{\partial \phi}{\partial X} - \frac{\partial \psi}{\partial Z}, \qquad w = \frac{\partial \phi}{\partial Z} + \frac{\partial \psi}{\partial X} \qquad (7.4.4)$$

where ϕ and ψ satisfy

$$v_p^2 \nabla^2 \phi = \frac{\partial^2 \phi}{\partial t^2}, \qquad v_s^2 \nabla^2 \psi = \frac{\partial^2 \psi}{\partial t^2} \qquad (7.4.5)$$

with $\nabla^2 = \partial^2/\partial X^2 + \partial^2/\partial Z^2$. The second component of the displacement field is given by

$$v = \frac{\partial \psi_1}{\partial Z} - \frac{\partial \psi_3}{\partial X}$$

where ψ_1 and ψ_3 satisfy $(7.4.2)_2$. However, the single equation

$$v_s^2 \nabla^2 v = \frac{\partial^2 v}{\partial t^2} \tag{7.4.6}$$

is more convenient when discussing the second component v.

One example in which ϕ and ψ are independent of Y is when the direction of propagation of a plane wave is in the XZ-plane. We take the Z-axis to be vertically downwards and $\mathbf{N} = (l, 0, n)$. Then

$$\phi = \phi\left(t - \frac{Xl + Zn}{v_p}\right), \qquad \psi = \psi\left(t - \frac{Xl + Zn}{v_s}\right)$$

$$v = v\left(t - \frac{Xl + Zn}{v_s}\right) \tag{7.4.7}$$

As in the general decomposition (7.4.1), ϕ is associated with a P-wave. Since ψ generates the displacement component in the XZ-plane which is the vertical plane containing the direction of propagation, it is associated with the SV-component of the shear wave, and v is associated with the SH-component.

Example 7.4.1

Using (7.4.4) and (7.4.7), show that

$$\mathbf{u} \cdot \mathbf{N} = -\frac{1}{v_p} \phi'\left(t - \frac{Xl + Zn}{v_p}\right)$$

$$\mathbf{u} \times \mathbf{N} = \left(vn, -\frac{1}{v_s} \psi'\left(t - \frac{Xl + Zn}{v_s}\right), -vl\right)$$

This example shows in particular that $\mathbf{u} \cdot \mathbf{N} = 0$ when $\phi = \text{constant}$, and $\mathbf{u} \times \mathbf{N} = \mathbf{0}$ when $v = 0$ and $\psi = \text{constant}$. In other words, for a P-wave the displacement is perpendicular to the wavefront since $\mathbf{u} \times \mathbf{N} = \mathbf{0}$ and $\mathbf{u} \cdot \mathbf{N} \neq 0$, whereas for an S-wave the displacement is in the plane of the wavefront.

In the following sections we wish to discuss boundary conditions on the planes $Z = \text{constant}$, and so we now calculate S_{ZX},

S_{ZY}, and S_{ZZ} in terms of ϕ, ψ, and v. Substituting (7.4.4) into (4.3.3) and using (7.3.3), we find

$$S_{ZX} = \rho_0 v_s^2 \left(2 \frac{\partial^2 \phi}{\partial X \partial Z} + \frac{\partial^2 \psi}{\partial X^2} - \frac{\partial^2 \psi}{\partial Z^2} \right)$$

$$S_{ZY} = \mu \frac{\partial v}{\partial Z} \qquad\qquad\qquad (7.4.8)$$

$$S_{ZZ} = \rho_0 (v_p^2 - 2v_s^2) \frac{\partial^2 \phi}{\partial X^2} + \rho_0 v_p^2 \frac{\partial^2 \phi}{\partial Z^2} + 2\rho_0 v_s^2 \frac{\partial^2 \psi}{\partial X \partial Z}$$

Example 7.4.2

Derive the relations (7.4.8).

7.5 Reflection of P-waves

When an earthquake disturbance reaches the interface of two different elastic materials, the incident wave is, in general, both reflected and refracted. Here we consider the simpler case of an elastic material occupying the region $Z \geqslant 0$ with the boundary $Z = 0$ free of applied traction. We consider first a P-wave encountering the surface at an acute angle of incidence θ, the Y-axis being taken perpendicular to the direction of propagation (see Fig. 7.3). Then $\mathbf{N} = (\sin \theta, 0, -\cos \theta)$, and the potential function

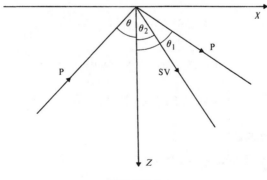

Fig. 7.3

ϕ_{in} representing this disturbance takes the form (cf. equation (7.2.4))

$$\phi_{in} = A \exp \left\{ i\omega \left(t - \frac{X \sin \theta - Z \cos \theta}{v_p} \right) \right\} \tag{7.5.1}$$

where A is a constant. We consider first the possibility that the reflected disturbance consists only of a P-wave propagating in a direction specified by $\mathbf{N} = (\sin \theta_1, 0, \cos \theta_1)$. This disturbance is then represented by the potential function

$$\phi_{ref} = B \exp \left\{ i\omega \left(t - \frac{X \sin \theta_1 + Z \cos \theta_1}{v_p} \right) \right\} \tag{7.5.2}$$

where B is a constant. If these are the only two waves present, the total disturbance at any point is represented by $\phi = \phi_{in} + \phi_{ref}$. Clearly, ϕ satisfies the wave equation (7.4.5)$_1$. Since we require the boundary $Z = 0$ to be traction-free, it follows from (7.4.8) that

$$2\rho_0 v_s^2 \frac{\partial^2 \phi}{\partial X \partial Z} = 0, \qquad \rho_0 (v_p^2 - 2v_s^2) \frac{\partial^2 \phi}{\partial X^2} + \rho_0 v_p^2 \frac{\partial^2 \phi}{\partial Z^2} = 0 \tag{7.5.3}$$

on $Z = 0$. Substituting $\phi = \phi_{in} + \phi_{ref}$ into (7.5.3), we obtain

$$\begin{aligned} A \sin 2\theta \exp & \left\{ i\omega \left(t - \frac{X \sin \theta}{v_p} \right) \right\} \\ & - B \sin 2\theta_1 \exp \left\{ i\omega \left(t - \frac{X \sin \theta_1}{v_p} \right) \right\} = 0 \\ (v_p^2 - 2v_s^2) & \left[A \sin^2 \theta \exp \left\{ i\omega \left(t - \frac{X \sin \theta}{v_p} \right) \right\} \right. \\ & \left. + B \sin^2 \theta_1 \exp \left\{ i\omega \left(t - \frac{X \sin \theta_1}{v_p} \right) \right\} \right] \\ + v_p^2 & \left[A \cos^2 \theta \exp \left\{ i\omega \left(t - \frac{X \sin \theta}{v_p} \right) \right\} \right. \\ & \left. + B \cos^2 \theta_1 \exp \left\{ i\omega \left(t - \frac{X \sin \theta_1}{v_p} \right) \right\} \right] = 0 \end{aligned} \tag{7.5.4}$$

Example 7.5.1

If functions f and g satisfy

$$f(x + \alpha y) = g(x + \beta y) \tag{7.5.5}$$

for all x and y, where α and β are constants, show that f and g are constant, or $\alpha = \beta$, and in either case $f = g$.

Since (7.5.5) holds for all x and y it must hold, in particular, for $y = 0$, so that $f(x) = g(x)$. Differentiating both sides of (7.5.5) partially with respect to y, and evaluating at $y = 0$, we obtain

$$\alpha f'(x) = \beta g'(x)$$

Since $f(x) = g(x)$, we must have either $f'(x) = g'(x) = 0$ and f and g are constant, or $\alpha = \beta$.

Since equations (7.5.4) hold for all X and t, it follows, using Example 7.5.1, that either $A = B = 0$ or

$$\sin \theta = \sin \theta_1 \tag{7.5.6}$$

For possible non-trivial solutions we consider equations (7.5.4) in the two situations: $\sin 2\theta = 0$, and $\sin 2\theta \neq 0$. Using (7.5.6) we see that if $\sin 2\theta = 0$ then $\sin 2\theta_1 = 0$ and $(7.5.4)_1$ is satisfied identically. Using (7.5.6) again, we see that if $\sin 2\theta = 0$ then either $\sin \theta = \sin \theta_1 = 0$, or $\cos \theta = \cos \theta_1 = 0$. In view of (7.3.5), $v_p^2 - 2v_s^2 > 0$ and so $(7.5.4)_2$ implies $A = -B$. Remembering that A and B are complex, we have $B = A e^{i\pi}$ and $|B/A| = 1$, so that the amplitude of the reflected wave is equal to that of the incident wave, but the phase of the incident and reflected waves differ by π.

Using (7.5.1) and (7.5.2) we see that at *grazing incidence* $(\theta = \pi/2)$, $\phi = \phi_{\text{in}} + \phi_{\text{ref}} = 0$, so that there is no disturbance. At *normal incidence* $(\theta = 0)$,

$$\phi = A\left[\exp\left\{i\omega\left(t + \frac{Z}{v_p}\right)\right\} - \exp\left\{i\omega\left(t - \frac{Z}{v_p}\right)\right\}\right]$$
$$= 2iA \exp(i\omega t) \sin\left(\frac{\omega Z}{v_p}\right) \tag{7.5.7}$$

which represents a *standing wave*.

Suppose now that $\sin 2\theta \neq 0$, then from $(7.5.4)_1$ we have

$$B = A$$

and $(7.5.4)_2$ reduces to

$$A(v_p^2 - 2v_s^2 \sin^2 \theta) = 0 \tag{7.5.8}$$

Since $v_p^2 > 2v_s^2$ for all known compressible materials, no non-trivial solution of (7.5.8) exists for any real value of θ. Hence, for a P-wave incident at an angle θ such that $\sin 2\theta \neq 0$, the traction-free boundary conditions cannot be satisfied by a disturbance in the form of a reflected P-wave only.

In view of the previous analysis we consider the possibility that an incident P-wave gives rise to a reflected disturbance consisting of both P- and S-waves. We take ϕ_{in} and ϕ_{ref} given by (7.5.1) and (7.5.2) and consider an S-wave reflected at an angle θ_2. This disturbance is represented by

$$\psi_{ref} = C \exp\left\{i\omega\left(t - \frac{X \sin\theta_2 + Z \cos\theta_2}{v_s}\right)\right\} \qquad (7.5.9)$$

$$v = D \exp\left\{i\omega\left(t - \frac{X \sin\theta_2 + Z \cos\theta_2}{v_s}\right)\right\} \qquad (7.5.10)$$

where C and D are constants. Since these satisfy $(7.4.5)_2$ and (7.4.6), it remains to consider the traction-free boundary conditions. Using (7.4.8) these become

$$2\frac{\partial^2 \phi}{\partial X \partial Z} + \frac{\partial^2 \psi}{\partial X^2} - \frac{\partial^2 \psi}{\partial Z^2} = 0$$

$$\frac{\partial v}{\partial Z} = 0 \qquad (7.5.11)$$

$$(v_p^2 - 2v_s^2)\frac{\partial^2 \phi}{\partial X^2} + v_p^2 \frac{\partial^2 \phi}{\partial Z^2} + 2v_s^2 \frac{\partial^2 \psi}{\partial X \partial Z} = 0$$

on $Z = 0$. On substituting (7.5.1), (7.5.2), (7.5.9), and (7.5.10) into $(7.5.11)_{1,3}$ we obtain

$$Aq^2 \sin 2\theta \exp\left\{i\omega\left(t - \frac{X \sin\theta}{v_p}\right)\right\}$$

$$- Bq^2 \sin 2\theta_1 \exp\left\{i\omega\left(t - \frac{X \sin\theta_1}{v_p}\right)\right\}$$

$$+ C \cos 2\theta_2 \exp\left\{i\omega\left(t - \frac{X \sin\theta_2}{v_s}\right)\right\} = 0$$

$$A(1 - 2q^2 \sin^2\theta) \exp\left\{i\omega\left(t - \frac{X \sin\theta}{v_p}\right)\right\} \qquad (7.5.12)$$

$$+ B(1 - 2q^2 \sin^2\theta_1) \exp\left\{i\omega\left(t - \frac{X \sin\theta_1}{v_p}\right)\right\}$$

$$+ C \sin 2\theta_2 \exp\left\{i\omega\left(t - \frac{X \sin\theta_2}{v_s}\right)\right\} = 0$$

where $q = v_s/v_p$. As before, these equations must hold for all X and t, so that we must have

$$\frac{\sin \theta}{v_p} = \frac{\sin \theta_1}{v_p} = \frac{\sin \theta_2}{v_s} \qquad (7.5.13)$$

and, restricting θ, θ_1, and θ_2 to be acute, it follows that

$$\theta_1 = \theta, \qquad \theta_2 < \theta \qquad (7.5.14)$$

Hence, the P-wave is reflected at the angle of incidence, and the S-wave is reflected at an angle less than the angle of incidence. Since $\theta \leqslant \pi/2$, $\cos \theta_2 \neq 0$, and so from $(7.5.11)_2$ we see that D must be zero. There is therefore no reflected SH-wave; the reflected S-wave has only SV-components. The amplitudes and phases of the reflected waves can be found by solving (7.5.12). At some angles of incidence one of the reflected waves is not present. For example, in agreement with the earlier analysis, when $\theta = 0$ there is no reflected SV-wave, whereas in a Poisson material for which $\lambda = \mu$, when $\theta = \pi/3$ the reflected P-wave is absent (see Example 7.5.2). However, this situation cannot arise for all values of v_p/v_s. For example, making use of the value $v_p/v_s = 1.788$, corresponding to $v_p = 7.17$ km/sec, $v_s = 4.01$ km/sec, Walker (1919) found that there is no real value of θ for which C vanishes.

Example 7.5.2

Using (7.5.12), show that there is no reflected P-wave for values of θ satisfying

$$2q^3 \sin \theta \sin 2\theta (1 - q^2 \sin^2 \theta)^{\frac{1}{2}} = (1 - 2q^2 \sin^2 \theta)^2$$

For a Poisson material show that one root of this equation is $\theta = \pi/3$.

Example 7.5.3

Find the displacement components on the surface $Z = 0$ resulting from an incident P-wave. [Leave your answer in complex form in terms of A, B, and C.]

Example 7.5.4

The region $Z \geqslant 0$ is occupied by an elastic material, the surface $Z = 0$ being free of applied traction. An SH-wave travelling in the medium encounters the free surface at an angle of incidence θ, the direction of incidence being perpendicular to the Y-axis. Assuming that the reflected disturbance consists of an SH-wave at an angle θ_1, show that $\theta = \theta_1$, and that the amplitude and phase of the incident and reflected waves are equal.

7.6   Reflection of SV-waves

As in the case of an incident P-wave, to satisfy the traction-free boundary condition on $Z = 0$ for an SV-wave incident at an arbitrary acute angle θ, it is necessary to have both a reflected SV-wave at an angle of reflection θ_1, and a reflected P-wave at an angle of reflection θ_2, where

$$\theta_1 = \theta, \qquad \sin\theta_2 = (v_p/v_s)\sin\theta \qquad (7.6.1)$$

A reflected SH-wave is not possible.

As in the previous case, there are particular values of θ in some materials for which the reflected SV-wave disappears, for example, in a Poisson material one such value is $\theta = \pi/6$. For grazing and normal incidence a reflected SV-wave alone satisfies the boundary conditions. In fact, in the former case, the disturbance vanishes entirely while in the latter case a standing wave is set up. The detailed analysis for an incident SV-wave is similar to that of the previous section and is not presented here (cf. Example 8 at the end of the chapter). However, we mention here how this case differs in one important aspect from the corresponding case of an incident P-wave.

From (7.6.1) we see that only when θ satisfies

$$0 \le \theta \le \sin^{-1}(v_s/v_p)$$

is the angle θ_2 real. If $\sin\theta > v_s/v_p$, then $\sin\theta_2 > 1$ and θ_2 is complex. In particular, $\cos\theta_2 = \pm ib'$ where b' is a positive real number giving

$$\phi_{\text{ref}} = Ce^{-bZ}\exp\left\{i\omega\left(t - \frac{X}{v_p'}\right)\right\}$$

where $b = \omega b'/v_p$ and $v_p' = v_p/\sin\theta_2 < v_p$. Here the negative square root has been taken so that the displacement remains bounded as $Z \to \infty$. This potential represents a disturbance which decays exponentially with distance from the boundary, and for this reason it is called a surface disturbance or surface wave (we consider two types of such wave in more detail in Section 7.8).

Thus for an incident SV-wave there are two possibilities depending on the size of the angle of incidence. If $\theta \leqslant \sin^{-1}(v_s/v_p)$, in general there is a reflected P-wave and a reflected SV-wave. If $\theta > \sin^{-1}(v_s/v_p)$, there is in general a reflected SV-wave and a surface disturbance propagating with a speed which is less than the speed of propagation of P-waves.

7.7 Reflection and refraction of plane harmonic waves

Suppose now that the region $Z > 0$, denoted by M, is occupied by one elastic material, and the region $Z < 0$, denoted by M', is occupied by a different elastic material. We assume that on the interface $Z = 0$ both the stress and the displacement are continuous. Waves may be reflected and refracted at such an interface. Since the analysis for the three possible types of incident wave is more complicated than that considered in the previous sections, we do not consider the problem of reflection and refraction in detail. The interested reader is referred to Ewing, Jardetzky, and Press (1957), Chapter 3. However, to illustrate how the boundary conditions are applied, we consider briefly the simplest case of an SH-wave incident through the medium M at an acute angle θ.

Let θ_1 and θ_2 be as shown in Fig. 7.4. Then, in M,

$$v = A \exp\left\{i\omega\left(t - \frac{X \sin\theta - Z \cos\theta}{v_s}\right)\right\}$$
$$+ B \exp\left\{i\omega\left(t - \frac{X \sin\theta_1 + Z \cos\theta_1}{v_s}\right)\right\} \qquad (7.7.1)$$

and in M',

$$v' = C \exp\left\{i\omega\left(t - \frac{X \sin\theta_2 - Z \cos\theta_2}{v'_s}\right)\right\} \qquad (7.7.2)$$

where $v'_s = (\mu'/\rho'_0)^{\frac{1}{2}}$ denotes the speed of S-waves in M', and A, B, and C are constants.

Since the displacement and stress must be continuous across $Z = 0$, we must have

$$v = v', \qquad \mu \frac{\partial v}{\partial Z} = \mu' \frac{\partial v'}{\partial Z} \qquad (7.7.3)$$

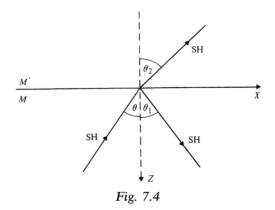

Fig. 7.4

on $Z = 0$. Substituting (7.7.1) and (7.7.2) into (7.7.3), we obtain

$$A \exp\left\{i\omega\left(t - \frac{X \sin \theta}{v_s}\right)\right\} + B \exp\left\{i\omega\left(t - \frac{X \sin \theta_1}{v_s}\right)\right\}$$

$$= C \exp\left\{i\omega\left(t - \frac{X \sin \theta_2}{v_s'}\right)\right\}$$

$$\mu\left[A \frac{\cos \theta}{v_s} \exp\left\{i\omega\left(t - \frac{X \sin \theta}{v_s}\right)\right\}\right. \tag{7.7.4}$$

$$\left. - B \frac{\cos \theta_1}{v_s} \exp\left\{i\omega\left(t - \frac{X \sin \theta_1}{v_s}\right)\right\}\right]$$

$$= \mu' C \frac{\cos \theta_2}{v_s'} \exp\left\{i\omega\left(t - \frac{X \sin \theta_2}{v_s'}\right)\right\}$$

Since these equations must hold for all X and t, it follows that

$$\frac{\sin \theta}{v_s} = \frac{\sin \theta_1}{v_s} = \frac{\sin \theta_2}{v_s'}$$

giving

$$\theta_1 = \theta, \qquad \sin \theta_2 = (v_s'/v_s) \sin \theta \tag{7.7.5}$$

Equations (7.7.4) then reduce to

$$A + B = C, \qquad \mu(A - B) \cos \theta = (\mu' v_s/v_s') C \cos \theta_2 \tag{7.7.6}$$

If $v_s' < v_s$ then from $(7.7.5)_2$ we see that $\theta_2 < \theta$ so that θ_2 is always real. On the other hand, if $v_s < v_s'$ there are two possibilities depending on the size of θ. If $0 < \theta \leqslant \sin^{-1}(v_s/v_s')$, then θ_2

is real, and in general there is both a reflected and a refracted SH-wave. However, if $\theta > \sin^{-1}(v_s/v_s')$, from reasoning similar to that given in Section 7.6, instead of a refracted SH-wave there is a surface disturbance.

7.8 Rayleigh waves

Following an earthquake, not only P- and S-waves are generated but, as mentioned earlier, surface waves can be detected. We now consider these waves in more detail.

Rayleigh (1885) set out 'to investigate the behaviour of waves upon the plane free surface of a semi-infinite homogeneous isotropic elastic solid, their character being such that the disturbance is confined to a superficial region, of thickness comparable with the wave length'. Since that time such waves have been called *Rayleigh waves*.

Suppose that the elastic material occupies the region $Z \geqslant 0$, and that the boundary $Z = 0$ is traction-free. A feature of Rayleigh waves is that the displacement lies in the planes $Y = $ constant so that $v = 0$. We therefore seek solutions of the form

$$\phi = F(Z) \exp\left\{i\omega\left(t - \frac{X}{V}\right)\right\}, \qquad \psi = G(Z) \exp\left\{i\omega\left(t - \frac{X}{V}\right)\right\} \quad (7.8.1)$$

where F and G are functions to be determined and $V \neq 0$ is assumed to be a real constant. Since $Z = 0$ is taken to be traction-free, we require

$$S_{ZX} = S_{ZY} = S_{ZZ} = 0 \qquad \text{on } Z = 0 \qquad (7.8.2)$$

Since we are seeking waves which are confined largely to a region near the surface, we suppose that

$$F(Z) \to 0, \qquad G(Z) \to 0 \qquad \text{as } Z \to \infty \qquad (7.8.3)$$

Substituting (7.8.1) into (7.4.5), we obtain

$$F''(Z) = \frac{\omega^2}{V^2}\left(1 - \frac{V^2}{v_p^2}\right)F(Z), \qquad G''(Z) = \frac{\omega^2}{V^2}\left(1 - \frac{V^2}{v_s^2}\right)G(Z) \quad (7.8.4)$$

which can be solved for the unknown functions F and G. If either $1 - (V/v_p)^2 < 0$ or $1 - (V/v_s)^2 < 0$, these equations give periodic solutions which would not satisfy (7.8.3). If $V = v_s$ then $1 - (V/v_p)^2 > 0$ since $v_s < v_p$, and (7.8.3) can be satisfied by taking

$G(Z) \equiv 0$. However, to satisfy $(7.8.2)_1$ and $(7.8.3)_1$, $F(Z)$ would also have to be identically zero and we obtain the trivial solution in which all the displacement components vanish. We see therefore that solutions of the form $(7.8.1)$ can represent surface waves only if

$$1 - (V/v_p)^2 > 0, \qquad 1 - (V/v_s)^2 > 0$$

and these inequalities hold provided

$$V^2 < v_s^2 \tag{7.8.5}$$

Assuming that V satisfies $(7.8.5)$, two pairs of linearly independent solutions of $(7.8.4)$ are $e^{\pm rZ}$ and $e^{\pm sZ}$, where the real numbers r, s are given by

$$r = \frac{\omega}{V}\left(1 - \frac{V^2}{v_p^2}\right)^{\frac{1}{2}} > 0, \qquad s = \frac{\omega}{V}\left(1 - \frac{V^2}{v_s^2}\right)^{\frac{1}{2}} > 0 \tag{7.8.6}$$

To satisfy $(7.8.3)$ we take the negative exponents, and the associated forms for ϕ and ψ are

$$\phi = A e^{-rZ} \exp\left\{i\omega\left(t - \frac{X}{V}\right)\right\}, \qquad \psi = B e^{-sZ} \exp\left\{i\omega\left(t - \frac{X}{V}\right)\right\} \tag{7.8.7}$$

where A and B are constants. It remains to satisfy $(7.8.2)$ and confirm $(7.8.5)$. Under our assumptions S_{ZY} is identically zero, so that $(7.8.2)$ are satisfied provided $S_{ZX} = S_{ZZ} = 0$ on $Z = 0$. Using $(7.4.8)_{1,3}$ and $(7.8.7)$, $(7.8.2)$ reduce to

$$2i\left(1 - \frac{V^2}{v_p^2}\right)^{\frac{1}{2}} A - \left(2 - \frac{V^2}{v_s^2}\right)B = 0$$

$$\left(2 - \frac{V^2}{v_s^2}\right)A + 2i\left(1 - \frac{V^2}{v_s^2}\right)^{\frac{1}{2}} B = 0 \tag{7.8.8}$$

For these equations to have a non-trivial solution, the determinant of coefficients must vanish. That is,

$$\left(2 - \frac{V^2}{v_s^2}\right)^2 = 4\left(1 - \frac{V^2}{v_s^2}\right)^{\frac{1}{2}}\left(1 - \frac{V^2}{v_p^2}\right)^{\frac{1}{2}} \tag{7.8.9}$$

Putting

$$p = (V/v_s)^2, \qquad q = v_s/v_p < 1 \tag{7.8.10}$$

and squaring both sides of (7.8.9), we obtain

$$(2-p)^4 = 16(1-p)(1-q^2p)$$

or

$$p\{p^3 - 8p^2 + 8(3-2q^2)p - 16(1-q^2)\} = 0$$

Hence, apart from the root $p = 0$ corresponding to the trivial solution $V = 0$, we require

$$f(p) \equiv p^3 - 8p^2 + 8(3-2q^2)p - 16(1-q^2) = 0 \qquad (7.8.11)$$

This is called the *Rayleigh equation*. Since $q^2 = (1-2\nu)/2(1-\nu)$, the roots of (7.8.11) depend only on Poisson's ratio. For a Rayleigh wave to exist, we require $0 < V < v_s$, or $0 < p < 1$. We show that (7.8.11) has one, and only one, real root in the interval $(0, 1)$. There is therefore just one solution of the assumed form which represents a Rayleigh wave.

Now

$$f(0) = -16(1-q^2) < 0, \quad \text{and} \quad f(1) = 1$$

showing that (7.8.11) has either one, or three, real roots in $(0, 1)$. If there were three real roots, $f''(p) = 2(3p - 8)$ would vanish in $(0, 1)$. Since this is not the case, we conclude that there is one real root p_0 in this interval, and hence one Rayleigh wave exists propagating with speed $V = v_s\sqrt{p_0} < v_s$.

The values of V/v_s for some typical values of ν are given in Table 7.2.

Table 7.2

ν	V/v_s
0.1	0.89
0.2	0.91
0.3	0.93
0.4	0.94

For a Poisson material $v_s/v_p = 1/\sqrt{3}$, so that $q^2 = 1/3$ and (7.8.11) reduces to

$$3p^3 - 24p^2 + 56p - 32 = 0$$

with roots $p = 4$, $2 \pm 2\sqrt{3}/3$. Only the root $2 - 2\sqrt{3}/3$ corresponds to a Rayleigh wave so that

$$V = 0.9194v_s$$

Having established in general that there is a value of V satisfying (7.8.11), equations (7.8.8) can be solved and hence, using (7.8.7) the displacement components can be found (see Example 12 at the end of the chapter). During the passage of the wave, the surface points can be shown to describe elliptical paths in the retrograde sense, the major axis of the ellipse being perpendicular to the surface $Z = 0$.

7.9   Love waves

If Rayleigh waves were the only type of surface wave, the horizontal component of displacement would be in the direction of propagation. However, surface waves with horizontal displacements perpendicular to the direction of propagation have been noticed on the records of actual earthquakes. We now consider whether surface waves of this type are possible within the theory. To this end we consider the possibility of solutions of the form

$$u = 0, \quad v = U(Z) \exp\left\{ i\omega\left(t - \frac{X}{V}\right)\right\}, \quad w = 0 \qquad (7.9.1)$$

satisfying (7.8.2), and

$$U(Z) \to 0 \quad \text{as} \quad Z \to \infty \qquad (7.9.2)$$

Recalling that the equations of motion reduce to (7.4.6) so that U satisfies $(7.8.4)_2$ and using the condition (7.9.2), we find

$$U(Z) = Ce^{-sZ}$$

where C is a constant and s is given by $(7.8.6)_2$. Under our assumptions, S_{ZX} and S_{ZZ} vanish identically, so that (7.8.2) are satisfied provided $S_{ZY} = 0$ on $Z = 0$. Using $(7.4.8)_2$ we see that this implies $C = 0$, so that waves of the type (7.9.1) cannot propagate along a traction-free boundary.

However, Love (1911) suggested that waves of this nature could be accounted for by assuming that the surface layer of the Earth differs from the interior. He considered a layer of thickness h overlying a deep layer. Let the Z-axis be vertically downwards, the X-axis in the direction of propagation, and the outer layer occupy the region $-h \leqslant Z \leqslant 0$. Suppose that elastic material in the region $Z > 0$ has shear modulus μ' and shear wave speed v'_s while the outer layer has shear modulus μ and shear wave speed v_s. We consider the possibility of solutions in which $u = w = 0$ in both media while v takes the form

$$
v = \begin{cases} U(Z) \exp\left\{ i\omega\left(t - \dfrac{X}{V}\right)\right\}, & -h \leqslant Z \leqslant 0 \\[2mm] W(Z) \exp\left\{ i\omega\left(t - \dfrac{X}{V}\right)\right\}, & Z \geqslant 0 \end{cases} \tag{7.9.3}
$$

We need to determine the functions U and W, and the constant V so that (7.9.3) satisfy (7.4.6) and appropriate interface and boundary conditions. Since we are considering waves whose amplitudes diminish to zero with distance from the interface, we also require that

$$
W(Z) \to 0 \qquad \text{as } Z \to \infty \tag{7.9.4}
$$

As in Section 7.7, we assume that the displacement and stress are continuous across $Z = 0$. This means that

$$
U(0) = W(0), \qquad \mu \frac{dU}{dZ} = \mu' \frac{dW}{dZ} \qquad \text{on } Z = 0 \tag{7.9.5}
$$

Finally, we assume that the surface $Z = -h$ is traction-free. Since the stresses S'_{ZX}, S'_{ZZ} associated with the displacement field (7.9.3)$_1$ vanish identically, this condition is satisfied if

$$
S'_{ZY} = 0 \qquad \text{on } Z = -h \tag{7.9.6}
$$

All primed quantities refer to the region of the layer $-h \leqslant Z \leqslant 0$.

Example 7.9.1

Assuming that $v_s < V < v'_s$ show that for a solution of the form (7.9.3) to satisfy (7.4.6) and (7.9.4),

$$
U(Z) = A \sin \sigma Z + B \cos \sigma Z, \qquad W(Z) = C e^{-\sigma' Z} \tag{7.9.7}
$$

where A, B, and C are real arbitrary constants, and

$$
\sigma = \frac{\omega}{V}\left(\frac{V^2}{v_s^2} - 1\right)^{\frac{1}{2}}, \qquad \sigma' = \frac{\omega}{V}\left(1 - \frac{V^2}{v_s'^2}\right)^{\frac{1}{2}} \tag{7.9.8}
$$

Example 7.9.2

Show that for (7.9.7) to be a non-trivial solution which satisfies (7.9.5) and (7.9.6), V must satisfy the equation

$$\mu'\left(1-\frac{V^2}{v_s'^2}\right)^{\frac{1}{2}} = \mu\left(\frac{V^2}{v_s^2}-1\right)^{\frac{1}{2}}\tan\left\{\frac{\omega}{V}\left(\frac{V^2}{v_s^2}-1\right)^{\frac{1}{2}}h\right\} \tag{7.9.9}$$

Equation (7.9.9) determines the speed V of Love waves. We see that, in contrast with the other wave motions considered in this chapter, the speed of propagation depends on the frequency ω. Waves with this property are said to be *dispersive*. The detailed analysis of (7.9.9) is beyond the scope of this volume and we refer the interested reader to Ewing, Jardetzky, and Press (1957) pp. 205–210.

The following examples show that, if the inequality $v_s < V < v_s'$ does not hold, then (7.9.5) and (7.9.6) cannot be satisfied.

Example 7.9.3

If $v_s' < V$, show that it is not possible to determine a function $W(Z)$ such that (7.9.3) is a non-trivial solution of the equations of motion satisfying (7.9.4).

Example 7.9.4

If $V < v_s' < v_s$, show that for non-trivial solutions of the form (7.9.3) to satisfy (7.9.4) and (7.9.5), V must satisfy the equation

$$e^{-2\sigma_2 h} = (\mu\sigma_2 + \mu'\sigma')/(\mu\sigma_2 - \mu'\sigma') \tag{7.9.10}$$

provided $\mu\sigma_2 \neq \mu'\sigma'$, where

$$\sigma_2 = \frac{\omega}{V}\left(1-\frac{V^2}{v_s^2}\right)^{\frac{1}{2}}$$

Show that the right-hand side of (7.9.10) is either negative or greater than 1.

It now follows that (7.9.10) cannot be satisfied since $0 < e^{-2\sigma_2 h} < 1$ for $0 < \sigma_2 h$. When $\mu\sigma_2 = \mu'\sigma'$, only the trivial solution satisfies all the conditions.

Examples 7

1. Show that, under the conditions of spherical symmetry (that is, when the displacement vector is everywhere directed radially from

the origin O, and depends only upon t and the distance $R = (X_K X_K)^{\frac{1}{2}}$ from O), the equations of motion of an elastic solid reduce to

$$\frac{\partial^2 u}{\partial t^2} = v_p^2 \left(\frac{\partial^2 u}{\partial R^2} + \frac{2}{R} \frac{\partial u}{\partial R} - \frac{2}{R^2} u \right)$$

where u is the radial displacement. Show that this equation is satisfied by

$$u = \frac{\partial}{\partial R} \left(\frac{\chi}{R} \right)$$

where $\chi(R, t)$ satisfies the equation

$$\frac{\partial^2 \chi}{\partial t^2} = v_p^2 \frac{\partial^2 \chi}{\partial R^2} \qquad (*)$$

Find the general solution of $(*)$ and interpret it.

2. A spherical cavity $0 \leqslant R \leqslant a$ in an elastic solid which is initially at rest is filled with explosive which is detonated at time $t = 0$. Suppose that this produces a pressure $P(t)$ on the internal surface $R = a$, where

$$P(t) = \begin{cases} P_0, & t \geqslant 0 \\ 0, & t < 0 \end{cases}$$

P_0 being constant. Assuming that as a result of the explosion an outgoing spherical wave is generated in the elastic solid represented by

$$\chi = f\{t - (R - a)/v_p\}$$

show that the radial displacement is given by

$$u = -\frac{1}{R^2} f\left\{ t - \frac{(R-a)}{v_p} \right\} - \frac{1}{R v_p} f'\left\{ t - \frac{(R-a)}{v_p} \right\}$$

Deduce that $f(t)$ satisfies the differential equation

$$\frac{\rho_0}{a} f''(t) + \frac{4\mu}{a^2 v_p} f'(t) + \frac{4\mu}{a^3} f(t) = \begin{cases} -P_0, & t \geqslant 0 \\ 0, & t < 0 \end{cases}$$

Hence, show that $f = 0$ for $t < 0$, and for $t \geqslant 0$

$$f = -\frac{a^3}{4\mu} P_0 + e^{-2q^2 t/\tau} \left\{ A \sin\left(\frac{\gamma t}{\tau} \right) + B \cos\left(\frac{\gamma t}{\tau} \right) \right\}$$

where

$$q = v_s/v_p, \quad \gamma = (1 - 2\nu)^{\frac{1}{2}}/(1 - \nu), \quad \tau = a/v_p$$

and A and B are constants.

[You may assume that

$$S_{RR} = (\lambda + 2\mu)\frac{\partial u}{\partial R} + 2\lambda\frac{u}{R}\bigg]$$

Note: By imposing the condition that, at the wavefront $R = a + v_p t$, the material is continuous so that $u = 0$, values of A and B can be obtained and the complete solution found. This problem, as well as that in which the pressure on the boundary is variable, have received a great deal of attention in an attempt to gain insight into the origin of earthquake phenomena and the waves generated by explosions.

3. The region $Z \geq 0$ is occupied by elastic material. At time $t = 0$ the material is in equilibrium in its reference configuration. For times $t > 0$ the surface $Z = 0$ is subjected to uniformly distributed time-dependent tractions giving rise to the boundary conditions

$$S_{ZX} = T\cos\theta\sin\omega t, \quad S_{ZY} = 0, \quad S_{ZZ} = T\sin\theta\sin\omega t \quad \text{on } Z = 0$$

where T and ω are positive constants, and θ is a fixed acute angle. Assuming that body force is absent, and that the wave motion excited in the body propagates in the positive Z-direction and depends only upon Z and t, find the components of displacement.

4. An elastic material occupies the region $0 \leq Z \leq h$. The plane $Z = h$ is kept fixed, and the motion of every point of the plane $Z = 0$ is described by

$$u = \varepsilon\cos\omega t, \quad v = 0, \quad w = \varepsilon\sin\omega t$$

where ε and ω are constants. Assuming that the motion is everywhere described by

$$u = f(Z)\cos\omega t, \quad v = 0, \quad w = g(Z)\sin\omega t$$

show that f and g satisfy the differential equations

$$f''(Z) + (\omega/v_s)^2 f(Z) = 0, \quad g''(Z) + (\omega/v_p)^2 g(Z) = 0$$

Hence, derive the solutions

$$u = \varepsilon \cos \omega t \sin \{\omega(h - Z)/v_s\}/\sin (\omega h/v_s)$$
$$w = \varepsilon \sin \omega t \sin \{\omega(h - Z)/v_p\}/\sin (\omega h/v_p)$$

and find the shear stress on the plane $Z = h$.

5. Assuming that body force is absent, show that a displacement field of the form

$$u = -Y\psi(Z, t), \quad v = X\psi(Z, t), \quad w = 0 \tag{1}$$

satisfies the equations of motion provided that

$$v_s^2 \frac{\partial^2 \psi}{\partial Z^2} = \frac{\partial^2 \psi}{\partial t^2} \tag{2}$$

A semi-infinite elastic cylinder occupies the region $0 \le X^2 + Y^2 \le a^2$, $Z \ge 0$. At time $t = 0$ the cylinder is in equilibrium in its undeformed configuration. For times $t > 0$ the curved surface remains traction-free, but the end face $Z = 0$ is rotated about its centre through an infinitesimal angle $f(t)$. Find the displacement field throughout the cylinder, and show that the couple acting on the end section has magnitude $(\pi \mu a^4/2v_s)f'(t)$.

6. An elastic cylinder of radius a and length l, occupying the region $0 \le X^2 + Y^2 \le a^2$, $0 \le Z \le l$, is set into torsional oscillations by rotating the end face $Z = 0$ about its centre through an angle $\tau \sin \omega t$, τ ($\ll 1$) and ω being positive constants. The end face $Z = l$ and the curved surface are traction-free. Show that the boundary conditions and the equations of motion are satisfied by the displacement field (1) of Example 5 provided that ψ satisfies equation (2) of Example 5 and

$$\psi(0, t) = \tau \sin \omega t, \quad \left(\frac{\partial \psi}{\partial Z}\right)_{Z=l} = 0$$

By considering normal-mode solutions of equation (2) of the form

$$\psi(Z, t) = F(Z) \sin \omega t$$

determine the function F which satisfies the conditions of the problem, and show that the magnitude of the couple which must be applied to the end face $Z = 0$ is $\frac{1}{2}\pi \rho_0 v_s \omega \tau a^4 \tan (\omega l/v_s) \sin \omega t$.

Note: Examples 5 and 6 illustrate the different wave motions which arise in an infinite cylinder and one of finite length. In the former a transverse wave propagates along the cylinder with speed v_s. The waveform can be determined when the rotation of the end section, or the form of the applied couple, is known. The associated displacement field may be written in cylindrical coordinates in the form $\mathbf{u} = Rf(t - Z/v_s)\mathbf{e}_\theta$. In the cylinder of finite length, the wave propagates with speed v_s but, on reaching the end $Z = l$, reflection occurs so that waves travel in both the positive and negative Z-directions, combining to form a standing wave.

7. The displacement field associated with a plane harmonic wave of constant amplitude propagating with speed V in the X-direction can be written in the form

$$\mathbf{u} = \mathbf{A} \exp \left\{ i\omega \left(t - \frac{X}{V} \right) \right\}$$

Show that this is a non-trivial solution of the equations of motion if either $V = v_p$ or $V = v_s$. Show further that: (i) when $V = v_p$ the displacement is in the X-direction and curl $\mathbf{u} = \mathbf{0}$; and (ii) when $V = v_s$ the displacement lies in the YZ-plane and div $\mathbf{u} = 0$.

8. Elastic material occupies the region $Z \geqslant 0$, the surface $Z = 0$ being free of applied traction. An SV-wave travelling in the medium encounters the free surface at an acute angle of incidence θ, the direction of propagation being perpendicular to the Y-axis. Assuming that the reflected disturbance consists of a P-wave at an acute angle θ_2 and an SV-wave at an acute angle θ_1, show that the boundary conditions can be satisfied provided that $\theta_1 = \theta$, and $\theta_2 > \theta$.
   Deduce that, if $0 < \sin \theta < q$ $(q = v_s/v_p)$, there is no reflected SV-wave if θ satisfies

$$2 \sin \theta \sin 2\theta (q^2 - \sin^2 \theta)^{\frac{1}{2}} = \cos^2 2\theta$$

For a Poisson material show that one root of this equation is $\theta = \pi/6$.

9. Elastic material occupies the region $Z \geqslant 0$, and the surface $Z = 0$ is in contact with a lubricated rigid plate which prevents motion

normal to the interface but exerts no tangential stress. A plane
P-wave is incident at the boundary $Z = 0$ at an acute angle of
incidence θ so that

$$\phi_{in} = f\left(t - \frac{X \sin \theta - Z \cos \theta}{v_p}\right)$$

Assuming that the reflected disturbance is represented by

$$\phi_{ref} = g\left(t - \frac{X \sin \theta_1 + Z \cos \theta_1}{v_p}\right)$$

$$\psi_{ref} = h\left(t - \frac{X \sin \theta_2 + Z \cos \theta_2}{v_s}\right)$$

where g and h are functions to be determined, show that there is
no reflected shear wave. Find g in terms of f and the angle of
reflection of the P-wave.

To satisfy the prescribed boundary conditions, the normal
stress on $Z = 0$ must be compressive, otherwise the plate and the
elastic material would separate. Show that this normal stress is
compressive provided that $f''(\xi) \leqslant 0$ for all relevant values of ξ.

10. Reconsider the situation described in Example 9 when the
incident wave is an SV-wave at an angle of incidence θ represented by

$$\psi_{in} = f\left(t - \frac{X \sin \theta - Z \cos \theta}{v_s}\right)$$

and discuss the nature of the reflected disturbance. Show that the
normal stress on $Z = 0$ is compressive provided $f''(\xi) \geqslant 0$ for all
relevant ξ.

11. An elastic plate occupies the region $X \geqslant 0$, $-\frac{1}{2}h \leqslant Z \leqslant \frac{1}{2}h$, and
its major faces $Z = \pm\frac{1}{2}h$ are traction-free. A motion is produced
in the plate by applying to the edge $X = 0$ a periodic traction
$S \sin (\pi Z/h) \sin \omega t$ in the Y-direction, where S and ω are positive
constants. Assuming that the displacement field is of the form

$$u = w = 0, \qquad v = v(X, Z, t)$$

show that if $\omega > \pi v_s/h$ the motion consists of a progressive wave
travelling in the X-direction, and find the speed of propagation.

12. If the constants A and B in (7.8.8) are expressed in the form

$$A = a\mathrm{e}^{\mathrm{i}\alpha}, \qquad B = b\mathrm{e}^{\mathrm{i}\beta}$$

where a, b, α, β are real, show that $\alpha = \beta - \frac{1}{2}\pi$ and

$$\frac{a}{b} = \frac{2 - V^2/v_s^2}{2(1 - V^2/v_p^2)^{\frac{1}{2}}}$$

Show also that, on $Z = 0$, the displacement components are

$$u = \frac{a\omega V}{2v_s^2} \sin\left\{\omega\left(t - \frac{X}{V}\right) + \alpha\right\}, \qquad w = \frac{b\omega V}{2v_s^2} \cos\left\{\omega\left(t - \frac{X}{V}\right) + \alpha\right\}$$

Deduce that the surface points describe elliptical paths in which the major axis of the ellipse is perpendicular to the undeformed surface. Show that the particles on the crests of the wave move in the direction opposite to that of propagation.

Answers

Chapter 1

(1.5.1) If $\lambda_i > 1$ the deformation represents an extension in the i-direction; if $0 < \lambda_i < 1$ it represents a contraction in this direction.

(1.5.2) αD

(1.5.5)

(a) Simple extension in the 1-direction combined with simple shearing of planes $X_2 = $ constant.

(b) A combination of the shearing of planes $X_2 = $ constant and $X_3 = $ constant.

(1.11.3)

$$T'_{11} = T_{11} \cos^2 \theta + T_{22} \sin^2 \theta + 2T_{12} \sin \theta \cos \theta$$

$$T'_{22} = T_{11} \sin^2 \theta + T_{22} \cos^2 \theta - 2T_{12} \sin \theta \cos \theta$$

$$T'_{12} = (T_{22} - T_{11}) \sin \theta \cos \theta + T_{12}(\cos^2 \theta - \sin^2 \theta)$$

$$T'_{13} = T_{13} \cos \theta + T_{23} \sin \theta$$

$$T'_{23} = -T_{13} \sin \theta + T_{23} \cos \theta$$

$$T'_{33} = T_{33}$$

(1.14.1)

$$T_{11} = S_{11} + \kappa S_{21}, \ T_{22} = S_{22}, \ T_{33} = S_{33}$$

$$T_{12} = S_{12} + \kappa S_{22} = S_{21}, \ T_{13} = S_{13} + \kappa S_{23} = S_{31}$$

$$T_{23} = S_{23} = S_{32}$$

S_{11}, S_{12} are the components in the 1- and 2-directions of the force (measured per unit area of $\mathscr{C}_0$) acting on an element of the surface $x_1 = a + \kappa x_2$ in $\mathscr{C}_t$ which corresponds to an element of the surface $X_1 = a$ in $\mathscr{C}_0$.

(1.15.1)

$$(l_{ij}) = \begin{pmatrix} \cos \theta & \sin \theta & 0 \\ -\sin \theta & \cos \theta & 0 \\ 0 & 0 & 1 \end{pmatrix}$$

$$T_{rr} = T \cos^2 \theta, \ T_{r\theta} = T_{\theta r} = -\tfrac{1}{2}T \sin 2\theta, \ T_{\theta\theta} = T \sin^2 \theta$$

$$T_{\theta z} = T_{z\theta} = T_{zr} = T_{rz} = T_{zz} = 0$$

Examples 1

1. The motion represents a uniform rotation in which particles move in circles with angular speed Ω

$v_1 = -\Omega(X_1 \sin \Omega t + X_2 \cos \Omega t),\ v_2 = \Omega(X_1 \cos \Omega t - X_2 \sin \Omega t),$

$v_3 = 0$ (material description)

$v_1 = -\Omega x_2,\ v_2 = \Omega x_1,\ v_3 = 0$ (spatial description)

$f_1 = -\Omega^2(X_1 \cos \Omega t - X_2 \sin \Omega t),\ f_2 = -\Omega^2(X_1 \sin \Omega t + X_2 \cos \Omega t),$

$f_3 = 0$ (material description)

$f_1 = -\Omega^2 x_1,\ f_2 = -\Omega^2 x_2,\ f_3 = 0$ (spatial description)

2. $J = \alpha\beta\gamma,$

$v_1 = \dot{\alpha}X_1,\ v_2 = \dot{\beta}X_2,\ v_3 = \dot{\gamma}X_3$ (material description)

$v_1 = \dfrac{\dot{\alpha}}{\alpha}\,x_1,\ v_2 = \dfrac{\dot{\beta}}{\beta}\,x_2,\ v_3 = \dfrac{\dot{\gamma}}{\gamma}\,x_3$ (spatial description)

4. Principal axes are inclined to the 1,2-axes at an angle $\tfrac{1}{2}\tan^{-1}(2/\kappa)$.

5. The plane $X_2 = \beta$ is deformed into the plane $x_2 = \tan(\alpha\beta)x_1$.
$f(X_1) = (2X_1 + A)^{\frac{1}{2}}/\alpha^{\frac{1}{2}}$, where A is an arbitrary constant.

6. $dl = \{(1+\alpha)^2 + \beta^2 + 2\beta(1+\alpha)\sin 2\theta\}^{\frac{1}{2}}\,dL.$

7. Each particle in the plane $X_3 = $ constant is displaced a distance proportional to X_3 and to the distance $R = (X_1^2 + X_2^2)^{\frac{1}{2}}$ from the X_3-axis, the direction of the displacement being at right angles to the radius vector (X_1, X_2).

$$\mathbf{F} = \begin{pmatrix} 1 & -\tau X_3 & -\tau X_2 \\ \tau X_3 & 1 & \tau X_1 \\ 0 & 0 & 1 \end{pmatrix}$$

$$\mathbf{C} = \begin{pmatrix} 1+\tau^2 X_3^2 & 0 & -\tau X_2 + \tau^2 X_1 X_3 \\ 0 & 1+\tau^2 X_3^2 & \tau X_1 + \tau^2 X_2 X_3 \\ -\tau X_2 + \tau^2 X_1 X_3 & \tau X_1 + \tau^2 X_2 X_3 & 1+\tau^2(X_1^2+X_2^2) \end{pmatrix}$$

The deformation is not isochoric (unless $\tau = 0$). The cylinder deforms into the hyperboloid of revolution

$$x_1^2 + x_2^2 = a^2(1 + \tau^2 x_3^2)$$

9. $\mathbf{t} = 2(1, 1, 1),\ \nu = 4,\ \sigma = 2\sqrt{3}$. The principal stresses are $-2,\ 1,\ 4$. Direction cosines of the principal axes are

$$\frac{1}{\sqrt{2}}(0, -1, 1),\quad \frac{1}{\sqrt{3}}(-1, 1, 1),\quad \frac{1}{\sqrt{6}}(2, 1, 1)$$

respectively.

10. $\left(-\sqrt{\dfrac{2}{5}}, \dfrac{1}{\sqrt{10}}, \dfrac{1}{\sqrt{2}}\right)$

11. The components of the surface traction on each face are:

$$\mathbf{t}(\mathbf{e}_1)|_{x_1=a} = \left(\frac{-p(a^2-x_2^2)}{a^2}, \frac{2px_2}{a}, 0\right)$$

$$\mathbf{t}(-\mathbf{e}_1)|_{x_1=-a} = -\mathbf{t}(\mathbf{e}_1)|_{x_1=-a} = \left(\frac{p(a^2-x_2^2)}{a^2}, \frac{2px_2}{a}, 0\right)$$

$$\mathbf{t}(\mathbf{e}_2)|_{x_2=a} = \left(\frac{2px_1}{a}, \frac{p(x_1^2-a^2)}{a^2}, 0\right)$$

$$\mathbf{t}(-\mathbf{e}_2)|_{x_2=-a} = -\mathbf{t}(\mathbf{e}_2)|_{x_2=-a} = \left(\frac{2px_1}{a}, -\frac{p(x_1^2-a^2)}{a^2}, 0\right)$$

$$\mathbf{t}(\mathbf{e}_3)|_{x_3=h} = \mathbf{0} = \mathbf{t}(-\mathbf{e}_3)|_{x_3=-h}$$

12. Body force is $-\omega^2(x_1, x_2, 0)$.

13. $\mathbf{t}(\mathbf{n}) = \begin{cases} T\mathbf{n} & \text{if } \mathbf{n} \text{ is parallel to } \mathbf{a} \\ \mathbf{0} & \text{if } \mathbf{n} \text{ is perpendicular to } \mathbf{a} \end{cases}$.

14. $T(r) = A/r^2$, where A is an arbitrary constant.

Chapter 3

(3.10.1) The moments about the 1- and 2-directions are

$$\int_S x_2 T_{33}\, dS \quad \text{and} \quad \int_S x_1 T_{33}\, dS$$

respectively.

Examples 3

1. $\kappa < 1$

2. Shearing of the planes $X_2 = \text{constant}$ of amount κ together with stretches β in the 1- and 3-directions and α in the 2-direction.

5. $p = \dfrac{k}{m}\left\{\left(\dfrac{A}{a}\right)^{3m} - 1\right\}$

6. $p_0 = 2(C_1 - C_2) + 4\kappa^2 C_2 A^2 - 8\kappa^2(C_1 + C_2)A^2$. The resultant force is $8\pi\kappa(C_1 + C_2)A^2 L$.

7. $p = \frac{1}{2}\rho\Omega^2(x_1^2 + x_2^2) + p_0$, where from the boundary conditions

$$p_0 = -\frac{1}{2}\rho\Omega^2\frac{A^2}{\alpha} + \frac{2C_1}{\alpha} - 2C_2\alpha$$

8. $\mathbf{B} = \begin{pmatrix} \dfrac{x_1^2}{A^2 r^4} + A^2 x_2^2 & \left(\dfrac{1}{A^2 r^4} - A^2\right) x_1 x_2 & 0 \\[3ex] \left(\dfrac{1}{A^2 r^4} - A^2\right) x_1 x_2 & \dfrac{x_2^2}{A^2 r^4} + A^2 x_1^2 & 0 \\[3ex] 0 & 0 & 1 \end{pmatrix}$

$\mathbf{B}^{-1} = \begin{pmatrix} \dfrac{x_2^2}{A^2 r^4} + A^2 x_1^2 & \left(A^2 - \dfrac{1}{A^2 r^4}\right) x_1 x_2 & 0 \\[3ex] \left(A^2 - \dfrac{1}{A^2 r^4}\right) x_1 x_2 & \dfrac{x_1^2}{A^2 r^4} + A^2 x_2^2 & 0 \\[3ex] 0 & 0 & 1 \end{pmatrix}$

Chapter 4

(4.8.1)   Extension $= 0.3 \times 10^{-2}$ mm; decrease in diameter $= 0.3 \times 10^{-4}$ mm.

(4.8.2)   The angle of shear is 0.35×10^{-3} radians.

Examples 4

1. $\mathbf{s}(\mathbf{N}) = \dfrac{\lambda T}{\lambda + 2\mu} \mathbf{N}$;   yes.

3. $\mathbf{u} = \begin{cases} A_1 R \mathbf{e}_R, & R \le a \\ (A_2 R + B_2/R^2) \mathbf{e}_R, & a \le R \le b \end{cases}$

where

$$B_2 = \frac{-pa^3 b^3 (\kappa_2 - \kappa_1)}{\kappa_2(\kappa_1 + 4\mu_2) b^3 - 4\mu_2(\kappa_2 - \kappa_1) a^3}, \quad \kappa_i = 3\lambda_i + 2\mu_i,$$

$$A_2 = \frac{\kappa_1 + 4\mu_2}{a^3(\kappa_2 - \kappa_1)} B_2, \quad A_1 = \frac{4\mu_2 + \kappa_2}{a^3(\kappa_2 - \kappa_1)} B_2.$$

4. $r = a - \dfrac{\rho k a^3}{5(3\lambda + 2\mu)}$

7. (i)   Support on rigid plinth with profile $X_3 = -\dfrac{\rho_0 g \nu}{2E}(X_1^2 + X_2^2)$.

   (ii)   Apply a uniform pressure $\rho_0 g l$ over the cross-section.

Chapter 5

(5.5.2) $S_{XX} = S_{YY} = S_{ZZ} = 0$, $S_{XY} = S$

(5.6.1) $2\mu(u+iv) = (1-2\nu)Tz$, $2\mu(u+iv) = iS\bar{z}$

Examples 5

1. $w = \varepsilon \dfrac{R}{a} \cos \Theta$, $S_{RZ} = \mu \dfrac{\varepsilon}{a} \cos \Theta$

2. $A = -\dfrac{3Fa}{2b^3}$, $B = \dfrac{3F}{2b^3}$, $C = -\dfrac{3F}{4b}$

3. $A = Sa^2$, $B = \frac{1}{2}S$, $C = -\frac{1}{2}a^4 S$; $S_{\Theta\Theta} = -4S \sin 2\Theta$

4. $A = \dfrac{i\mu\alpha a^2}{2(1-\nu)(a^2-b^2)}$, $B = -\dfrac{2i\mu\alpha a^2 b^2}{a^2-b^2}$, $M = -\dfrac{4\pi\mu a^2 b^2 \alpha}{a^2-b^2}$

Chapter 6

(6.7.1) $\chi = \frac{1}{2}(a^2 - X^2 - Y^2)$

Examples 6

1. On $R = b$,

$S_{XZ} = -\mu\tau \sin \Theta(b - 2a \cos \Theta)$, $S_{YZ} = \mu\tau \cos \Theta(b - 2a \cos \Theta)$

   On $R = 2a \cos \Theta$,

$$S_{XZ} = -\mu\tau\{1 - (b^2/4a^2) \sec^2 \Theta\}a \sin 2\Theta$$
$$S_{YZ} = \mu\tau\{1 - (b^2/4a^2) \sec^2 \Theta\}a \cos 2\Theta$$

The total shear is greatest on the boundary $R = 2a \cos \Theta$ at $\Theta = 0$.

3. $M = \pi\mu\tau a^3 b^3 (1 - k^4)/(a^2 + b^2)$

Chapter 7

(7.5.3)

$$u = \left\{ -(A+B)\frac{\sin \theta}{v_p} + C\frac{(1 - q^2 \sin^2 \theta)^{\frac{1}{2}}}{v_s} \right\} i\omega \exp \left\{ i\omega\left(t - \frac{X \sin \theta}{v_p} \right) \right\}$$

$$v = 0, \; w = \left\{ (A-B)\frac{\cos \theta}{v_p} - C\frac{\sin \theta}{v_p} \right\} i\omega \exp \left\{ i\omega\left(t - \frac{X \sin \theta}{v_p} \right) \right\}$$

Examples 7

1. $\chi = f(t - R/v_p) + g(t + R/v_p)$, where f and g are arbitrary functions. The first term represents a spherical wave propagating away from the origin with speed v_p, while the second term represents a spherical wave travelling towards the origin with speed v_p.

3. $u = \begin{cases} \dfrac{T\cos\theta}{\rho_0 v_s \omega}\left[\cos\left\{\omega\left(t - \dfrac{Z}{v_p}\right)\right\} - 1\right], & t - \dfrac{Z}{v_s} \geq 0 \\ 0, & t - Z/v_s < 0 \end{cases}$

$v = 0$, $w = \begin{cases} \dfrac{T\sin\theta}{\rho_0 v_p \omega}\left[\cos\left\{\omega\left(t - \dfrac{Z}{v_p}\right)\right\} - 1\right], & t - \dfrac{Z}{v_p} \geq 0 \\ 0, & t - Z/v_p < 0 \end{cases}$

4. $S_{ZX} = -\rho_0 \omega v_s \varepsilon \cos \omega t \cos ec(\omega h/v_s)$

5. $\psi = \begin{cases} 0, & t \leq Z/v_s \\ f(t - Z/v_s), & t > Z/v_s \end{cases}$

6. $F = \tau\{\cos(\omega Z/v_s) + \tan(\omega l/v_s)\sin(\omega Z/v_s)\}$

9. $g = f$, $\quad \theta_1 = \theta$

10. For all angles of incidence there is only a reflected SV-wave at an angle of reflection equal to the angle of incidence.

11. The speed of propagation is $\omega h v_s/(h^2\omega^2 - \pi^2 v_s^2)^{\frac{1}{2}}$.

References and suggestions for further reading

Blatz, P. J. (1969) 'Application of large deformation theory to the thermo-mechanical behavior of rubberlike polymers – porous, unfilled, and filled', *Rheology–Theory and Applications*, vol. 5 (ed. F. R. Eirich), 1–55. Academic Press.

Blatz, P. J. & Ko, W. L. (1962) 'Application of finite elasticity theory to deformation of rubbery materials' *Trans. Soc. Rheology*, **6**, 223–51.

Bullen, K. E. (1963) *An Introduction to the Theory of Seismology* (3rd edn), Cambridge University Press.

Chadwick, P. (1976) *Continuum Mechanics, Concise Theory and Problems*, George Allen & Unwin.

England, A. H. (1971) *Complex Variable Methods in Elasticity*, Wiley.

Ericksen, J. L. (1955) 'Deformations possible in every compressible isotropic perfectly elastic material', *J. Math. Phys.*, **34**, 126–8.

Eringen, A. C. & Suhubi, E. S. (1975) *Elastodynamics*, vol. II *Linear Theory*, Academic Press.

Ewing, W. M., Jardetzky, W. S. & Press, F. (1957) *Elastic Waves in Layered Media*, McGraw-Hill.

Gent, A. N. & Rivlin, R. S. (1952) 'Experiments on the mechanics of rubber II: The torsion, inflation and extension of a tube', *Proc. Phys. Soc. Lond.*, **B65**, 487–501.

Green, A. E. & Adkins, J. E. (1970) *Large Elastic Deformations and Non-linear Continuum Mechanics* (2nd edn revised by A. E. Green), Oxford University Press.

Green, A. E. & Zerna, W. (1968) *Theoretical Elasticity* (2nd edn), Oxford University Press.

Gurtin, M. E. (1972) 'The linear theory of elasticity', *Handbuch der Physik* (ed. S. Flügge), vol. VIa/2, Springer-Verlag.

Hunter, S. C. (1976) *Mechanics of Continuous Media*, Ellis Horwood.

Jeffreys, H. (1976) *The Earth* (6th edn), Cambridge University Press.

Love, A. E. H. (1911) *Some Problems of Geodynamics*, Cambridge University Press.

Love, A. E. H. (1952) *A Treatise on the Mathematical Theory of Elasticity* (4th edn), Cambridge University Press.

Mooney, M. (1940) 'A theory of large elastic deformation', *J. Appl. Phys.*, **11**, 582–92.

Morse, P. M. & Feshbach, H. (1953) *Methods of Theoretical Physics*, McGraw-Hill.

Muskhelishvili, N. I. (1953) *Some Basic Problems of the Mathematical Theory of Elasticity*, P. Noordhoff.

Ogden, R. W. (1972a) 'Large deformation isotropic elasticity – on the correlation of theory and experiment for incompressible rubberlike solids', *Proc. R. Soc. Lond.*, **A326**, 565–84.

Ogden, R. W. (1972b) 'Large deformation isotropic elasticity – on the correlation of theory and experiment for compressible rubberlike solids', *Proc. R. Soc. Lond.*, **A328**, 567–83.

Oldham, R. D. (1900) 'On the propagation of earthquake motion to great distances', *Phil. Trans. R. Soc. Lond.*, **A194**, 135–74.

Phillips, E. G. (1957) *Functions of a Complex Variable* (8th edn), Longman.

Poynting, J. H. (1909) 'On pressure perpendicular to the shear planes in finite pure shears, and on the lengthening of loaded wires when twisted', *Proc. R. Soc.*, **A82**, 546–59.

Rayleigh, Lord (1885) 'On waves propagated along the plane surface of an elastic solid', *Proc. London Math. Soc.*, **17**, 4–11.

Rivlin, R. S. (1947) 'Torsion of a rubber cylinder', *J. Appl. Phys.*, **18**, 444–9.

Rivlin, R. S. (1949) 'Large elastic deformations of isotropic materials VI: Further results in the theory of torsion, shear and flexure', *Phil. Trans. R. Soc. Lond.*, **A242**, 173–95.

Rivlin, R. S. & Saunders, D. W. (1951) 'Large elastic deformations of isotropic materials VII: Experiments on the deformation of rubber', *Phil. Trans. R. Soc. Lond.*, **A243**, 251–88.

Sokolnikoff, I. S. (1956) *Mathematical Theory of Elasticity* (2nd edn), McGraw-Hill.

Spencer, A. J. M. (1970) 'The static theory of finite elasticity', *J. Inst. Maths Applics.*, **6**, 164–200.

Spencer, A. J. M. (1980) *Continuum Mechanics*, Longman.

Sternberg, E. (1960) 'On the integration of the equations of motion in the classical theory of elasticity', *Arch. Rational Mech. Anal.*, **6**, 34–50.

Treloar, L. R. G. (1948) 'Stresses and birefringence in rubber subjected to general homogeneous strain', *Proc. Phys. Soc.*, **60**, 135–44.

Treloar, L. R. G. (1958) *Physics of Rubber Elasticity*, Oxford University Press.

Treloar, L. R. G. (1973) 'The elasticity and related properties of rubbers', *Rep. Prog. Phys.*, **36**, 755–826.

Truesdell, C. & Noll, W. (1965) 'The non-linear field theories of mechanics', *Handbuch der Physik* (ed. S. Flügge), vol. III/3, Springer-Verlag.

Truesdell, C. & Toupin, R. A. (1960) 'The classical field theories', *Handbuch der Physik* (ed. S. Flügge), vol. III/1, Springer-Verlag.

Walker, G. W. (1919) 'Surface reflection of earthquake waves', *Phil. Trans. R. Soc. Lond.*, **A218**, 373–93.

Yeh, H. & Abrams, J. L. (1960) *Principles of Mechanics of Solids and Fluids*, vol. 1 *Particle and Rigid-body Mechanics*, McGraw-Hill.

Index

A CATALOG OF SELECTED
DOVER BOOKS
IN SCIENCE AND MATHEMATICS

Astronomy

BURNHAM'S CELESTIAL HANDBOOK, Robert Burnham, Jr. Thorough guide to the stars beyond our solar system. Exhaustive treatment. Alphabetical by constellation: Andromeda to Cetus in Vol. 1; Chamaeleon to Orion in Vol. 2; and Pavo to Vulpecula in Vol. 3. Hundreds of illustrations. Index in Vol. 3. 2,000pp. 6⅛ x 9¼.
Vol. I: 23567-X
Vol. II: 23568-8
Vol. III: 23673-0

EXPLORING THE MOON THROUGH BINOCULARS AND SMALL TELESCOPES, Ernest H. Cherrington, Jr. Informative, profusely illustrated guide to locating and identifying craters, rills, seas, mountains, other lunar features. Newly revised and updated with special section of new photos. Over 100 photos and diagrams. 240pp. 8¼ x 11. 24491-1

THE EXTRATERRESTRIAL LIFE DEBATE, 1750–1900, Michael J. Crowe. First detailed, scholarly study in English of the many ideas that developed from 1750 to 1900 regarding the existence of intelligent extraterrestrial life. Examines ideas of Kant, Herschel, Voltaire, Percival Lowell, many other scientists and thinkers. 16 illustrations. 704pp. 5⅜ x 8½. 40675-X

THEORIES OF THE WORLD FROM ANTIQUITY TO THE COPERNICAN REVOLUTION, Michael J. Crowe. Newly revised edition of an accessible, enlightening book recreates the change from an earth-centered to a sun-centered conception of the solar system. 242pp. 5⅜ x 8½. 41444-2

A HISTORY OF ASTRONOMY, A. Pannekoek. Well-balanced, carefully reasoned study covers such topics as Ptolemaic theory, work of Copernicus, Kepler, Newton, Eddington's work on stars, much more. Illustrated. References. 521pp. 5⅜ x 8½.
65994-1

A COMPLETE MANUAL OF AMATEUR ASTRONOMY: Tools and Techniques for Astronomical Observations, P. Clay Sherrod with Thomas L. Koed. Concise, highly readable book discusses: selecting, setting up and maintaining a telescope; amateur studies of the sun; lunar topography and occultations; observations of Mars, Jupiter, Saturn, the minor planets and the stars; an introduction to photoelectric photometry; more. 1981 ed. 124 figures. 26 halftones. 37 tables. 335pp. 6½ x 9¼.
42820-6

AMATEUR ASTRONOMER'S HANDBOOK, J. B. Sidgwick. Timeless, comprehensive coverage of telescopes, mirrors, lenses, mountings, telescope drives, micrometers, spectroscopes, more. 189 illustrations. 576pp. 5⅝ x 8¼. (Available in U.S. only.)
24034-7

STARS AND RELATIVITY, Ya. B. Zel'dovich and I. D. Novikov. Vol. 1 of *Relativistic Astrophysics* by famed Russian scientists. General relativity, properties of matter under astrophysical conditions, stars, and stellar systems. Deep physical insights, clear presentation. 1971 edition. References. 544pp. 5⅝ x 8¼. 69424-0

Chemistry

THE SCEPTICAL CHYMIST: The Classic 1661 Text, Robert Boyle. Boyle defines the term "element," asserting that all natural phenomena can be explained by the motion and organization of primary particles. 1911 ed. viii+232pp. 5⅜ x 8½.
42825-7

RADIOACTIVE SUBSTANCES, Marie Curie. Here is the celebrated scientist's doctoral thesis, the prelude to her receipt of the 1903 Nobel Prize. Curie discusses establishing atomic character of radioactivity found in compounds of uranium and thorium; extraction from pitchblende of polonium and radium; isolation of pure radium chloride; determination of atomic weight of radium; plus electric, photographic, luminous, heat, color effects of radioactivity. ii+94pp. 5⅜ x 8½.
42550-9

CHEMICAL MAGIC, Leonard A. Ford. Second Edition, Revised by E. Winston Grundmeier. Over 100 unusual stunts demonstrating cold fire, dust explosions, much more. Text explains scientific principles and stresses safety precautions. 128pp. 5⅜ x 8½.
67628-5

THE DEVELOPMENT OF MODERN CHEMISTRY, Aaron J. Ihde. Authoritative history of chemistry from ancient Greek theory to 20th-century innovation. Covers major chemists and their discoveries. 209 illustrations. 14 tables. Bibliographies. Indices. Appendices. 851pp. 5⅜ x 8½.
64235-6

CATALYSIS IN CHEMISTRY AND ENZYMOLOGY, William P. Jencks. Exceptionally clear coverage of mechanisms for catalysis, forces in aqueous solution, carbonyl- and acyl-group reactions, practical kinetics, more. 864pp. 5⅜ x 8½.
65460-5

ELEMENTS OF CHEMISTRY, Antoine Lavoisier. Monumental classic by founder of modern chemistry in remarkable reprint of rare 1790 Kerr translation. A must for every student of chemistry or the history of science. 539pp. 5⅜ x 8½. 64624-6

THE HISTORICAL BACKGROUND OF CHEMISTRY, Henry M. Leicester. Evolution of ideas, not individual biography. Concentrates on formulation of a coherent set of chemical laws. 260pp. 5⅜ x 8½.
61053-5

A SHORT HISTORY OF CHEMISTRY, J. R. Partington. Classic exposition explores origins of chemistry, alchemy, early medical chemistry, nature of atmosphere, theory of valency, laws and structure of atomic theory, much more. 428pp. 5⅜ x 8½. (Available in U.S. only.)
65977-1

GENERAL CHEMISTRY, Linus Pauling. Revised 3rd edition of classic first-year text by Nobel laureate. Atomic and molecular structure, quantum mechanics, statistical mechanics, thermodynamics correlated with descriptive chemistry. Problems. 992pp. 5⅜ x 8½.
65622-5

FROM ALCHEMY TO CHEMISTRY, John Read. Broad, humanistic treatment focuses on great figures of chemistry and ideas that revolutionized the science. 50 illustrations. 240pp. 5⅜ x 8½.
28690-8

Engineering

DE RE METALLICA, Georgius Agricola. The famous Hoover translation of greatest treatise on technological chemistry, engineering, geology, mining of early modern times (1556). All 289 original woodcuts. 638pp. 6¾ x 11. 60006-8

FUNDAMENTALS OF ASTRODYNAMICS, Roger Bate et al. Modern approach developed by U.S. Air Force Academy. Designed as a first course. Problems, exercises. Numerous illustrations. 455pp. 5⅜ x 8½. 60061-0

DYNAMICS OF FLUIDS IN POROUS MEDIA, Jacob Bear. For advanced students of ground water hydrology, soil mechanics and physics, drainage and irrigation engineering, and more. 335 illustrations. Exercises, with answers. 784pp. 6⅛ x 9¼. 65675-6

THEORY OF VISCOELASTICITY (Second Edition), Richard M. Christensen. Complete, consistent description of the linear theory of the viscoelastic behavior of materials. Problem-solving techniques discussed. 1982 edition. 29 figures. xiv+364pp. 6⅛ x 9¼. 42880-X

MECHANICS, J. P. Den Hartog. A classic introductory text or refresher. Hundreds of applications and design problems illuminate fundamentals of trusses, loaded beams and cables, etc. 334 answered problems. 462pp. 5⅜ x 8½. 60754-2

MECHANICAL VIBRATIONS, J. P. Den Hartog. Classic textbook offers lucid explanations and illustrative models, applying theories of vibrations to a variety of practical industrial engineering problems. Numerous figures. 233 problems, solutions. Appendix. Index. Preface. 436pp. 5⅜ x 8½. 64785-4

STRENGTH OF MATERIALS, J. P. Den Hartog. Full, clear treatment of basic material (tension, torsion, bending, etc.) plus advanced material on engineering methods, applications. 350 answered problems. 323pp. 5⅜ x 8½. 60755-0

A HISTORY OF MECHANICS, René Dugas. Monumental study of mechanical principles from antiquity to quantum mechanics. Contributions of ancient Greeks, Galileo, Leonardo, Kepler, Lagrange, many others. 671pp. 5⅜ x 8½. 65632-2

STABILITY THEORY AND ITS APPLICATIONS TO STRUCTURAL MECHANICS, Clive L. Dym. Self-contained text focuses on Koiter postbuckling analyses, with mathematical notions of stability of motion. Basing minimum energy principles for static stability upon dynamic concepts of stability of motion, it develops asymptotic buckling and postbuckling analyses from potential energy considerations, with applications to columns, plates, and arches. 1974 ed. 208pp. 5⅜ x 8½. 42541-X

METAL FATIGUE, N. E. Frost, K. J. Marsh, and L. P. Pook. Definitive, clearly written, and well-illustrated volume addresses all aspects of the subject, from the historical development of understanding metal fatigue to vital concepts of the cyclic stress that causes a crack to grow. Includes 7 appendixes. 544pp. 5⅜ x 8½. 40927-9

CATALOG OF DOVER BOOKS

ROCKETS, Robert Goddard. Two of the most significant publications in the history of rocketry and jet propulsion: "A Method of Reaching Extreme Altitudes" (1919) and "Liquid Propellant Rocket Development" (1936). 128pp. 5⅜ x 8½.   42537-1

STATISTICAL MECHANICS: Principles and Applications, Terrell L. Hill. Standard text covers fundamentals of statistical mechanics, applications to fluctuation theory, imperfect gases, distribution functions, more. 448pp. 5⅜ x 8½.   65390-0

ENGINEERING AND TECHNOLOGY 1650–1750: Illustrations and Texts from Original Sources, Martin Jensen. Highly readable text with more than 200 contemporary drawings and detailed engravings of engineering projects dealing with surveying, leveling, materials, hand tools, lifting equipment, transport and erection, piling, bailing, water supply, hydraulic engineering, and more. Among the specific projects outlined—transporting a 50-ton stone to the Louvre, erecting an obelisk, building timber locks, and dredging canals. 207pp. 8⅜ x 11¼.   42232-1

THE VARIATIONAL PRINCIPLES OF MECHANICS, Cornelius Lanczos. Graduate level coverage of calculus of variations, equations of motion, relativistic mechanics, more. First inexpensive paperbound edition of classic treatise. Index. Bibliography. 418pp. 5⅜ x 8½.   65067-7

PROTECTION OF ELECTRONIC CIRCUITS FROM OVERVOLTAGES, Ronald B. Standler. Five-part treatment presents practical rules and strategies for circuits designed to protect electronic systems from damage by transient overvoltages. 1989 ed. xxiv+434pp. 6⅛ x 9¼.   42552-5

ROTARY WING AERODYNAMICS, W. Z. Stepniewski. Clear, concise text covers aerodynamic phenomena of the rotor and offers guidelines for helicopter performance evaluation. Originally prepared for NASA. 537 figures. 640pp. 6⅛ x 9¼.   64647-5

INTRODUCTION TO SPACE DYNAMICS, William Tyrrell Thomson. Comprehensive, classic introduction to space-flight engineering for advanced undergraduate and graduate students. Includes vector algebra, kinematics, transformation of coordinates. Bibliography. Index. 352pp. 5⅜ x 8½.   65113-4

HISTORY OF STRENGTH OF MATERIALS, Stephen P. Timoshenko. Excellent historical survey of the strength of materials with many references to the theories of elasticity and structure. 245 figures. 452pp. 5⅜ x 8½.   61187-6

ANALYTICAL FRACTURE MECHANICS, David J. Unger. Self-contained text supplements standard fracture mechanics texts by focusing on analytical methods for determining crack-tip stress and strain fields. 336pp. 6⅛ x 9¼.   41737-9

STATISTICAL MECHANICS OF ELASTICITY, J. H. Weiner. Advanced, self-contained treatment illustrates general principles and elastic behavior of solids. Part 1, based on classical mechanics, studies thermoelastic behavior of crystalline and polymeric solids. Part 2, based on quantum mechanics, focuses on interatomic force laws, behavior of solids, and thermally activated processes. For students of physics and chemistry and for polymer physicists. 1983 ed. 96 figures. 496pp. 5⅜ x 8½.   42260-7

Mathematics

FUNCTIONAL ANALYSIS (Second Corrected Edition), George Bachman and Lawrence Narici. Excellent treatment of subject geared toward students with background in linear algebra, advanced calculus, physics, and engineering. Text covers introduction to inner-product spaces, normed, metric spaces, and topological spaces; complete orthonormal sets, the Hahn-Banach Theorem and its consequences, and many other related subjects. 1966 ed. 544pp. 6⅛ x 9¼. 40251-7

ASYMPTOTIC EXPANSIONS OF INTEGRALS, Norman Bleistein & Richard A. Handelsman. Best introduction to important field with applications in a variety of scientific disciplines. New preface. Problems. Diagrams. Tables. Bibliography. Index. 448pp. 5⅜ x 8½. 65082-0

VECTOR AND TENSOR ANALYSIS WITH APPLICATIONS, A. I. Borisenko and I. E. Tarapov. Concise introduction. Worked-out problems, solutions, exercises. 257pp. 5⅝ x 8¼. 63833-2

THE ABSOLUTE DIFFERENTIAL CALCULUS (CALCULUS OF TENSORS), Tullio Levi-Civita. Great 20th-century mathematician's classic work on material necessary for mathematical grasp of theory of relativity. 452pp. 5⅝ x 8¼. 63401-9

AN INTRODUCTION TO ORDINARY DIFFERENTIAL EQUATIONS, Earl A. Coddington. A thorough and systematic first course in elementary differential equations for undergraduates in mathematics and science, with many exercises and problems (with answers). Index. 304pp. 5⅜ x 8½. 65942-9

FOURIER SERIES AND ORTHOGONAL FUNCTIONS, Harry F. Davis. An incisive text combining theory and practical example to introduce Fourier series, orthogonal functions and applications of the Fourier method to boundary-value problems. 570 exercises. Answers and notes. 416pp. 5⅜ x 8½. 65973-9

COMPUTABILITY AND UNSOLVABILITY, Martin Davis. Classic graduate-level introduction to theory of computability, usually referred to as theory of recurrent functions. New preface and appendix. 288pp. 5⅜ x 8½. 61471-9

ASYMPTOTIC METHODS IN ANALYSIS, N. G. de Bruijn. An inexpensive, comprehensive guide to asymptotic methods–the pioneering work that teaches by explaining worked examples in detail. Index. 224pp. 5⅜ x 8½ 64221-6

APPLIED COMPLEX VARIABLES, John W. Dettman. Step-by-step coverage of fundamentals of analytic function theory–plus lucid exposition of five important applications: Potential Theory; Ordinary Differential Equations; Fourier Transforms; Laplace Transforms; Asymptotic Expansions. 66 figures. Exercises at chapter ends. 512pp. 5⅜ x 8½. 64670-X

INTRODUCTION TO LINEAR ALGEBRA AND DIFFERENTIAL EQUATIONS, John W. Dettman. Excellent text covers complex numbers, determinants, orthonormal bases, Laplace transforms, much more. Exercises with solutions. Undergraduate level. 416pp. 5⅜ x 8½. 65191-6

CALCULUS OF VARIATIONS WITH APPLICATIONS, George M. Ewing. Applications-oriented introduction to variational theory develops insight and promotes understanding of specialized books, research papers. Suitable for advanced undergraduate/graduate students as primary, supplementary text. 352pp. 5⅜ x 8½.
64856-7

COMPLEX VARIABLES, Francis J. Flanigan. Unusual approach, delaying complex algebra till harmonic functions have been analyzed from real variable viewpoint. Includes problems with answers. 364pp. 5⅜ x 8½. 61388-7

AN INTRODUCTION TO THE CALCULUS OF VARIATIONS, Charles Fox. Graduate-level text covers variations of an integral, isoperimetrical problems, least action, special relativity, approximations, more. References. 279pp. 5⅜ x 8½.
65499-0

COUNTEREXAMPLES IN ANALYSIS, Bernard R. Gelbaum and John M. H. Olmsted. These counterexamples deal mostly with the part of analysis known as "real variables." The first half covers the real number system, and the second half encompasses higher dimensions. 1962 edition. xxiv+198pp. 5⅜ x 8½. 42875-3

CATASTROPHE THEORY FOR SCIENTISTS AND ENGINEERS, Robert Gilmore. Advanced-level treatment describes mathematics of theory grounded in the work of Poincaré, R. Thom, other mathematicians. Also important applications to problems in mathematics, physics, chemistry, and engineering. 1981 edition. References. 28 tables. 397 black-and-white illustrations. xvii+666pp. 6⅛ x 9¼.
67539-4

INTRODUCTION TO DIFFERENCE EQUATIONS, Samuel Goldberg. Exceptionally clear exposition of important discipline with applications to sociology, psychology, economics. Many illustrative examples; over 250 problems. 260pp. 5⅜ x 8½.
65084-7

NUMERICAL METHODS FOR SCIENTISTS AND ENGINEERS, Richard Hamming. Classic text stresses frequency approach in coverage of algorithms, polynomial approximation, Fourier approximation, exponential approximation, other topics. Revised and enlarged 2nd edition. 721pp. 5⅜ x 8½. 65241-6

INTRODUCTION TO NUMERICAL ANALYSIS (2nd Edition), F. B. Hildebrand. Classic, fundamental treatment covers computation, approximation, interpolation, numerical differentiation and integration, other topics. 150 new problems. 669pp. 5⅜ x 8½. 65363-3

THREE PEARLS OF NUMBER THEORY, A. Y. Khinchin. Three compelling puzzles require proof of a basic law governing the world of numbers. Challenges concern van der Waerden's theorem, the Landau-Schnirelmann hypothesis and Mann's theorem, and a solution to Waring's problem. Solutions included. 64pp. 5⅜ x 8½.
40026-3

THE PHILOSOPHY OF MATHEMATICS: An Introductory Essay, Stephan Körner. Surveys the views of Plato, Aristotle, Leibniz & Kant concerning propositions and theories of applied and pure mathematics. Introduction. Two appendices. Index. 198pp. 5⅜ x 8½. 25048-2

CATALOG OF DOVER BOOKS

INTRODUCTORY REAL ANALYSIS, A.N. Kolmogorov, S. V. Fomin. Translated by Richard A. Silverman. Self-contained, evenly paced introduction to real and functional analysis. Some 350 problems. 403pp. 5⅜ x 8½. 61226-0

APPLIED ANALYSIS, Cornelius Lanczos. Classic work on analysis and design of finite processes for approximating solution of analytical problems. Algebraic equations, matrices, harmonic analysis, quadrature methods, more. 559pp. 5⅜ x 8½. 65656-X

AN INTRODUCTION TO ALGEBRAIC STRUCTURES, Joseph Landin. Superb self-contained text covers "abstract algebra": sets and numbers, theory of groups, theory of rings, much more. Numerous well-chosen examples, exercises. 247pp. 5⅜ x 8½. 65940-2

QUALITATIVE THEORY OF DIFFERENTIAL EQUATIONS, V. V. Nemytskii and V.V. Stepanov. Classic graduate-level text by two prominent Soviet mathematicians covers classical differential equations as well as topological dynamics and ergodic theory. Bibliographies. 523pp. 5⅜ x 8½. 65954-2

THEORY OF MATRICES, Sam Perlis. Outstanding text covering rank, nonsingularity and inverses in connection with the development of canonical matrices under the relation of equivalence, and without the intervention of determinants. Includes exercises. 237pp. 5⅜ x 8½. 66810-X

INTRODUCTION TO ANALYSIS, Maxwell Rosenlicht. Unusually clear, accessible coverage of set theory, real number system, metric spaces, continuous functions, Riemann integration, multiple integrals, more. Wide range of problems. Undergraduate level. Bibliography. 254pp. 5⅜ x 8½. 65038-3

MODERN NONLINEAR EQUATIONS, Thomas L. Saaty. Emphasizes practical solution of problems; covers seven types of equations. ". . . a welcome contribution to the existing literature. . . . "–*Math Reviews*. 490pp. 5⅜ x 8½. 64232-1

MATRICES AND LINEAR ALGEBRA, Hans Schneider and George Phillip Barker. Basic textbook covers theory of matrices and its applications to systems of linear equations and related topics such as determinants, eigenvalues, and differential equations. Numerous exercises. 432pp. 5⅜ x 8½. 66014-1

MATHEMATICS APPLIED TO CONTINUUM MECHANICS, Lee A. Segel. Analyzes models of fluid flow and solid deformation. For upper-level math, science, and engineering students. 608pp. 5⅜ x 8½. 65369-2

ELEMENTS OF REAL ANALYSIS, David A. Sprecher. Classic text covers fundamental concepts, real number system, point sets, functions of a real variable, Fourier series, much more. Over 500 exercises. 352pp. 5⅜ x 8½. 65385-4

SET THEORY AND LOGIC, Robert R. Stoll. Lucid introduction to unified theory of mathematical concepts. Set theory and logic seen as tools for conceptual understanding of real number system. 496pp. 5⅜ x 8¼. 63829-4

TENSOR CALCULUS, J.L. Synge and A. Schild. Widely used introductory text covers spaces and tensors, basic operations in Riemannian space, non-Riemannian spaces, etc. 324pp. 5⅜ x 8¼.   63612-7

ORDINARY DIFFERENTIAL EQUATIONS, Morris Tenenbaum and Harry Pollard. Exhaustive survey of ordinary differential equations for undergraduates in mathematics, engineering, science. Thorough analysis of theorems. Diagrams. Bibliography. Index. 818pp. 5⅜ x 8½.   64940-7

INTEGRAL EQUATIONS, F. G. Tricomi. Authoritative, well-written treatment of extremely useful mathematical tool with wide applications. Volterra Equations, Fredholm Equations, much more. Advanced undergraduate to graduate level. Exercises. Bibliography. 238pp. 5⅜ x 8½.   64828-1

FOURIER SERIES, Georgi P. Tolstov. Translated by Richard A. Silverman. A valuable addition to the literature on the subject, moving clearly from subject to subject and theorem to theorem. 107 problems, answers. 336pp. 5⅜ x 8½.   63317-9

INTRODUCTION TO MATHEMATICAL THINKING, Friedrich Waismann. Examinations of arithmetic, geometry, and theory of integers; rational and natural numbers; complete induction; limit and point of accumulation; remarkable curves; complex and hypercomplex numbers, more. 1959 ed. 27 figures. xii+260pp. 5⅜ x 8½.   42804-4

POPULAR LECTURES ON MATHEMATICAL LOGIC, Hao Wang. Noted logician's lucid treatment of historical developments, set theory, model theory, recursion theory and constructivism, proof theory, more. 3 appendixes. Bibliography. 1981 ed. ix+283pp. 5⅜ x 8½.   67632-3

CALCULUS OF VARIATIONS, Robert Weinstock. Basic introduction covering isoperimetric problems, theory of elasticity, quantum mechanics, electrostatics, etc. Exercises throughout. 326pp. 5⅜ x 8½.   63069-2

THE CONTINUUM: A Critical Examination of the Foundation of Analysis, Hermann Weyl. Classic of 20th-century foundational research deals with the conceptual problem posed by the continuum. 156pp. 5⅜ x 8½.   67982-9

CHALLENGING MATHEMATICAL PROBLEMS WITH ELEMENTARY SOLUTIONS, A. M. Yaglom and I. M. Yaglom. Over 170 challenging problems on probability theory, combinatorial analysis, points and lines, topology, convex polygons, many other topics. Solutions. Total of 445pp. 5⅜ x 8½. Two-vol. set.
Vol. I: 65536-9 Vol. II: 65537-7

INTRODUCTION TO PARTIAL DIFFERENTIAL EQUATIONS WITH APPLICATIONS, E. C. Zachmanoglou and Dale W. Thoe. Essentials of partial differential equations applied to common problems in engineering and the physical sciences. Problems and answers. 416pp. 5⅜ x 8½.   65251-3

THE THEORY OF GROUPS, Hans J. Zassenhaus. Well-written graduate-level text acquaints reader with group-theoretic methods and demonstrates their usefulness in mathematics. Axioms, the calculus of complexes, homomorphic mapping, p-group theory, more. 276pp. 5⅜ x 8½.   40922-8

Math–Decision Theory, Statistics, Probability

ELEMENTARY DECISION THEORY, Herman Chernoff and Lincoln E. Moses. Clear introduction to statistics and statistical theory covers data processing, probability and random variables, testing hypotheses, much more. Exercises. 364pp. 5⅜ x 8½.   65218-1

STATISTICS MANUAL, Edwin L. Crow et al. Comprehensive, practical collection of classical and modern methods prepared by U.S. Naval Ordnance Test Station. Stress on use. Basics of statistics assumed. 288pp. 5⅜ x 8½.   60599-X

SOME THEORY OF SAMPLING, William Edwards Deming. Analysis of the problems, theory, and design of sampling techniques for social scientists, industrial managers, and others who find statistics important at work. 61 tables. 90 figures. xvii +602pp. 5⅜ x 8½.   64684-X

LINEAR PROGRAMMING AND ECONOMIC ANALYSIS, Robert Dorfman, Paul A. Samuelson and Robert M. Solow. First comprehensive treatment of linear programming in standard economic analysis. Game theory, modern welfare economics, Leontief input-output, more. 525pp. 5⅜ x 8½.   65491-5

PROBABILITY: An Introduction, Samuel Goldberg. Excellent basic text covers set theory, probability theory for finite sample spaces, binomial theorem, much more. 360 problems. Bibliographies. 322pp. 5⅜ x 8½.   65252-1

GAMES AND DECISIONS: Introduction and Critical Survey, R. Duncan Luce and Howard Raiffa. Superb nontechnical introduction to game theory, primarily applied to social sciences. Utility theory, zero-sum games, n-person games, decision-making, much more. Bibliography. 509pp. 5⅜ x 8½.   65943-7

INTRODUCTION TO THE THEORY OF GAMES, J. C. C. McKinsey. This comprehensive overview of the mathematical theory of games illustrates applications to situations involving conflicts of interest, including economic, social, political, and military contexts. Appropriate for advanced undergraduate and graduate courses; advanced calculus a prerequisite. 1952 ed. x+372pp. 5⅜ x 8½.   42811-7

FIFTY CHALLENGING PROBLEMS IN PROBABILITY WITH SOLUTIONS, Frederick Mosteller. Remarkable puzzlers, graded in difficulty, illustrate elementary and advanced aspects of probability. Detailed solutions. 88pp. 5⅜ x 8½.   65355-2

PROBABILITY THEORY: A Concise Course, Y. A. Rozanov. Highly readable, self-contained introduction covers combination of events, dependent events, Bernoulli trials, etc. 148pp. 5⅜ x 8¼.   63544-9

STATISTICAL METHOD FROM THE VIEWPOINT OF QUALITY CONTROL, Walter A. Shewhart. Important text explains regulation of variables, uses of statistical control to achieve quality control in industry, agriculture, other areas. 192pp. 5⅜ x 8½.   65232-7

Math–Geometry and Topology

ELEMENTARY CONCEPTS OF TOPOLOGY, Paul Alexandroff. Elegant, intuitive approach to topology from set-theoretic topology to Betti groups; how concepts of topology are useful in math and physics. 25 figures. 57pp. $5\frac{3}{8} \times 8\frac{1}{2}$. 60747-X

COMBINATORIAL TOPOLOGY, P. S. Alexandrov. Clearly written, well-organized, three-part text begins by dealing with certain classic problems without using the formal techniques of homology theory and advances to the central concept, the Betti groups. Numerous detailed examples. 654pp. $5\frac{3}{8} \times 8\frac{1}{2}$. 40179-0

EXPERIMENTS IN TOPOLOGY, Stephen Barr. Classic, lively explanation of one of the byways of mathematics. Klein bottles, Moebius strips, projective planes, map coloring, problem of the Koenigsberg bridges, much more, described with clarity and wit. 43 figures. 210pp. $5\frac{3}{8} \times 8\frac{1}{2}$. 25933-1

CONFORMAL MAPPING ON RIEMANN SURFACES, Harvey Cohn. Lucid, insightful book presents ideal coverage of subject. 334 exercises make book perfect for self-study. 55 figures. 352pp. $5\frac{5}{8} \times 8\frac{1}{4}$. 64025-6

THE GEOMETRY OF RENÉ DESCARTES, René Descartes. The great work founded analytical geometry. Original French text, Descartes's own diagrams, together with definitive Smith-Latham translation. 244pp. $5\frac{3}{8} \times 8\frac{1}{2}$. 60068-8

PRACTICAL CONIC SECTIONS: The Geometric Properties of Ellipses, Parabolas and Hyperbolas, J. W. Downs. This text shows how to create ellipses, parabolas, and hyperbolas. It also presents historical background on their ancient origins and describes the reflective properties and roles of curves in design applications. 1993 ed. 98 figures. xii+100pp. $6\frac{1}{2} \times 9\frac{1}{4}$. 42876-1

THE THIRTEEN BOOKS OF EUCLID'S ELEMENTS, translated with introduction and commentary by Thomas L. Heath. Definitive edition. Textual and linguistic notes, mathematical analysis. 2,500 years of critical commentary. Unabridged. 1,414pp. $5\frac{3}{8} \times 8\frac{1}{2}$. Three-vol. set. Vol. I: 60088-2 Vol. II: 60089-0 Vol. III: 60090-4

GEOMETRY OF COMPLEX NUMBERS, Hans Schwerdtfeger. Illuminating, widely praised book on analytic geometry of circles, the Moebius transformation, and two-dimensional non-Euclidean geometries. 200pp. $5\frac{5}{8} \times 8\frac{1}{4}$. 63830-8

DIFFERENTIAL GEOMETRY, Heinrich W. Guggenheimer. Local differential geometry as an application of advanced calculus and linear algebra. Curvature, transformation groups, surfaces, more. Exercises. 62 figures. 378pp. $5\frac{3}{8} \times 8\frac{1}{2}$. 63433-7

CURVATURE AND HOMOLOGY: Enlarged Edition, Samuel I. Goldberg. Revised edition examines topology of differentiable manifolds; curvature, homology of Riemannian manifolds; compact Lie groups; complex manifolds; curvature, homology of Kaehler manifolds. New Preface. Four new appendixes. 416pp. $5\frac{3}{8} \times 8\frac{1}{2}$. 40207-X

History of Math

THE WORKS OF ARCHIMEDES, Archimedes (T. L. Heath, ed.). Topics include the famous problems of the ratio of the areas of a cylinder and an inscribed sphere; the measurement of a circle; the properties of conoids, spheroids, and spirals; and the quadrature of the parabola. Informative introduction. clxxxvi+326pp; supplement, 52pp. 5⅜ x 8½. 42084-1

A SHORT ACCOUNT OF THE HISTORY OF MATHEMATICS, W. W. Rouse Ball. One of clearest, most authoritative surveys from the Egyptians and Phoenicians through 19th-century figures such as Grassman, Galois, Riemann. Fourth edition. 522pp. 5⅜ x 8½. 20630-0

THE HISTORY OF THE CALCULUS AND ITS CONCEPTUAL DEVELOP-MENT, Carl B. Boyer. Origins in antiquity, medieval contributions, work of Newton, Leibniz, rigorous formulation. Treatment is verbal. 346pp. 5⅜ x 8½. 60509-4

THE HISTORICAL ROOTS OF ELEMENTARY MATHEMATICS, Lucas N. H. Bunt, Phillip S. Jones, and Jack D. Bedient. Fundamental underpinnings of modern arithmetic, algebra, geometry, and number systems derived from ancient civilizations. 320pp. 5⅜ x 8½. 25563-8

A HISTORY OF MATHEMATICAL NOTATIONS, Florian Cajori. This classic study notes the first appearance of a mathematical symbol and its origin, the competition it encountered, its spread among writers in different countries, its rise to popularity, its eventual decline or ultimate survival. Original 1929 two-volume edition presented here in one volume. xxviii+820pp. 5⅜ x 8½. 67766-4

GAMES, GODS & GAMBLING: A History of Probability and Statistical Ideas, F. N. David. Episodes from the lives of Galileo, Fermat, Pascal, and others illustrate this fascinating account of the roots of mathematics. Features thought-provoking references to classics, archaeology, biography, poetry. 1962 edition. 304pp. 5⅜ x 8½. (Available in U.S. only.) 40023-9

OF MEN AND NUMBERS: The Story of the Great Mathematicians, Jane Muir. Fascinating accounts of the lives and accomplishments of history's greatest mathematical minds–Pythagoras, Descartes, Euler, Pascal, Cantor, many more. Anecdotal, illuminating. 30 diagrams. Bibliography. 256pp. 5⅜ x 8½. 28973-7

HISTORY OF MATHEMATICS, David E. Smith. Nontechnical survey from ancient Greece and Orient to late 19th century; evolution of arithmetic, geometry, trigonometry, calculating devices, algebra, the calculus. 362 illustrations. 1,355pp. 5⅜ x 8½. Two-vol. set. Vol. I: 20429-4 Vol. II: 20430-8

A CONCISE HISTORY OF MATHEMATICS, Dirk J. Struik. The best brief history of mathematics. Stresses origins and covers every major figure from ancient Near East to 19th century. 41 illustrations. 195pp. 5⅜ x 8½. 60255-9

Physics

OPTICAL RESONANCE AND TWO-LEVEL ATOMS, L. Allen and J. H. Eberly. Clear, comprehensive introduction to basic principles behind all quantum optical resonance phenomena. 53 illustrations. Preface. Index. 256pp. 5⅜ x 8½. 65533-4

QUANTUM THEORY, David Bohm. This advanced undergraduate-level text presents the quantum theory in terms of qualitative and imaginative concepts, followed by specific applications worked out in mathematical detail. Preface. Index. 655pp. 5⅜ x 8½. 65969-0

ATOMIC PHYSICS: 8th edition, Max Born. Nobel laureate's lucid treatment of kinetic theory of gases, elementary particles, nuclear atom, wave-corpuscles, atomic structure and spectral lines, much more. Over 40 appendices, bibliography. 495pp. 5⅜ x 8½. 65984-4

A SOPHISTICATE'S PRIMER OF RELATIVITY, P. W. Bridgman. Geared toward readers already acquainted with special relativity, this book transcends the view of theory as a working tool to answer natural questions: What is a frame of reference? What is a "law of nature"? What is the role of the "observer"? Extensive treatment, written in terms accessible to those without a scientific background. 1983 ed. xlviii+172pp. 5⅜ x 8½. 42549-5

AN INTRODUCTION TO HAMILTONIAN OPTICS, H. A. Buchdahl. Detailed account of the Hamiltonian treatment of aberration theory in geometrical optics. Many classes of optical systems defined in terms of the symmetries they possess. Problems with detailed solutions. 1970 edition. xv+360pp. 5⅜ x 8½. 67597-1

PRIMER OF QUANTUM MECHANICS, Marvin Chester. Introductory text examines the classical quantum bead on a track: its state and representations; operator eigenvalues; harmonic oscillator and bound bead in a symmetric force field; and bead in a spherical shell. Other topics include spin, matrices, and the structure of quantum mechanics; the simplest atom; indistinguishable particles; and stationary-state perturbation theory. 1992 ed. xiv+314pp. 6⅛ x 9¼. 42878-8

LECTURES ON QUANTUM MECHANICS, Paul A. M. Dirac. Four concise, brilliant lectures on mathematical methods in quantum mechanics from Nobel Prize–winning quantum pioneer build on idea of visualizing quantum theory through the use of classical mechanics. 96pp. 5⅜ x 8½. 41713-1

THIRTY YEARS THAT SHOOK PHYSICS: The Story of Quantum Theory, George Gamow. Lucid, accessible introduction to influential theory of energy and matter. Careful explanations of Dirac's anti-particles, Bohr's model of the atom, much more. 12 plates. Numerous drawings. 240pp. 5⅜ x 8½. 24895-X

ELECTRONIC STRUCTURE AND THE PROPERTIES OF SOLIDS: The Physics of the Chemical Bond, Walter A. Harrison. Innovative text offers basic understanding of the electronic structure of covalent and ionic solids, simple metals, transition metals and their compounds. Problems. 1980 edition. 582pp. 6⅛ x 9¼. 66021-4

CATALOG OF DOVER BOOKS

HYDRODYNAMIC AND HYDROMAGNETIC STABILITY, S. Chandrasekhar. Lucid examination of the Rayleigh-Benard problem; clear coverage of the theory of instabilities causing convection. 704pp. 5⅜ x 8¼. 64071-X

INVESTIGATIONS ON THE THEORY OF THE BROWNIAN MOVEMENT, Albert Einstein. Five papers (1905–8) investigating dynamics of Brownian motion and evolving elementary theory. Notes by R. Fürth. 122pp. 5⅜ x 8½. 60304-0

THE PHYSICS OF WAVES, William C. Elmore and Mark A. Heald. Unique overview of classical wave theory. Acoustics, optics, electromagnetic radiation, more. Ideal as classroom text or for self-study. Problems. 477pp. 5⅜ x 8½. 64926-1

PHYSICAL PRINCIPLES OF THE QUANTUM THEORY, Werner Heisenberg. Nobel Laureate discusses quantum theory, uncertainty, wave mechanics, work of Dirac, Schroedinger, Compton, Wilson, Einstein, etc. 184pp. 5⅜ x 8½. 60113-7

ATOMIC SPECTRA AND ATOMIC STRUCTURE, Gerhard Herzberg. One of best introductions; especially for specialist in other fields. Treatment is physical rather than mathematical. 80 illustrations. 257pp. 5⅜ x 8½. 60115-3

AN INTRODUCTION TO STATISTICAL THERMODYNAMICS, Terrell L. Hill. Excellent basic text offers wide-ranging coverage of quantum statistical mechanics, systems of interacting molecules, quantum statistics, more. 523pp. 5⅜ x 8½. 65242-4

THEORETICAL PHYSICS, Georg Joos, with Ira M. Freeman. Classic overview covers essential math, mechanics, electromagnetic theory, thermodynamics, quantum mechanics, nuclear physics, other topics. xxiii+885pp. 5⅜ x 8½. 65227-0

PROBLEMS AND SOLUTIONS IN QUANTUM CHEMISTRY AND PHYSICS, Charles S. Johnson, Jr. and Lee G. Pedersen. Unusually varied problems, detailed solutions in coverage of quantum mechanics, wave mechanics, angular momentum, molecular spectroscopy, more. 280 problems, 139 supplementary exercises. 430pp. 6½ x 9¼. 65236-X

THEORETICAL SOLID STATE PHYSICS, Vol. I: Perfect Lattices in Equilibrium; Vol. II: Non-Equilibrium and Disorder, William Jones and Norman H. March. Monumental reference work covers fundamental theory of equilibrium properties of perfect crystalline solids, non-equilibrium properties, defects and disordered systems. Total of 1,301pp. 5⅜ x 8½. Vol. I: 65015-4 Vol. II: 65016-2

WHAT IS RELATIVITY? L. D. Landau and G. B. Rumer. Written by a Nobel Prize physicist and his distinguished colleague, this compelling book explains the special theory of relativity to readers with no scientific background, using such familiar objects as trains, rulers, and clocks. 1960 ed. vi+72pp. 23 b/w illustrations. 5⅜ x 8½. 42806-0 $6.95

A TREATISE ON ELECTRICITY AND MAGNETISM, James Clerk Maxwell. Important foundation work of modern physics. Brings to final form Maxwell's theory of electromagnetism and rigorously derives his general equations of field theory. 1,084pp. 5⅜ x 8½. Two-vol. set. Vol. I: 60636-8 Vol. II: 60637-6

QUANTUM MECHANICS: Principles and Formalism, Roy McWeeny. Graduate student–oriented volume develops subject as fundamental discipline, opening with review of origins of Schrödinger's equations and vector spaces. Focusing on main principles of quantum mechanics and their immediate consequences, it concludes with final generalizations covering alternative "languages" or representations. 1972 ed. 15 figures. xi+155pp. 5⅜ x 8½. 42829-X

INTRODUCTION TO QUANTUM MECHANICS WITH APPLICATIONS TO CHEMISTRY, Linus Pauling & E. Bright Wilson, Jr. Classic undergraduate text by Nobel Prize winner applies quantum mechanics to chemical and physical problems. Numerous tables and figures enhance the text. Chapter bibliographies. Appendices. Index. 468pp. 5⅜ x 8½. 64871-0

METHODS OF THERMODYNAMICS, Howard Reiss. Outstanding text focuses on physical technique of thermodynamics, typical problem areas of understanding, and significance and use of thermodynamic potential. 1965 edition. 238pp. 5⅜ x 8½. 69445-3

TENSOR ANALYSIS FOR PHYSICISTS, J. A. Schouten. Concise exposition of the mathematical basis of tensor analysis, integrated with well-chosen physical examples of the theory. Exercises. Index. Bibliography. 289pp. 5⅜ x 8½. 65582-2

THE ELECTROMAGNETIC FIELD, Albert Shadowitz. Comprehensive undergraduate text covers basics of electric and magnetic fields, builds up to electromagnetic theory. Also related topics, including relativity. Over 900 problems. 768pp. 5⅜ x 8¼. 65660-8

GREAT EXPERIMENTS IN PHYSICS: Firsthand Accounts from Galileo to Einstein, Morris H. Shamos (ed.). 25 crucial discoveries: Newton's laws of motion, Chadwick's study of the neutron, Hertz on electromagnetic waves, more. Original accounts clearly annotated. 370pp. 5⅜ x 8½. 25346-5

RELATIVITY, THERMODYNAMICS AND COSMOLOGY, Richard C. Tolman. Landmark study extends thermodynamics to special, general relativity; also applications of relativistic mechanics, thermodynamics to cosmological models. 501pp. 5⅜ x 8½. 65383-8

STATISTICAL PHYSICS, Gregory H. Wannier. Classic text combines thermodynamics, statistical mechanics, and kinetic theory in one unified presentation of thermal physics. Problems with solutions. Bibliography. 532pp. 5⅜ x 8½. 65401-X

Paperbound unless otherwise indicated. Available at your book dealer, online at **www.doverpublications.com**, or by writing to Dept. GI, Dover Publications, Inc., 31 East 2nd Street, Mineola, NY 11501. For current price information or for free catalogs (please indicate field of interest), write to Dover Publications or log on to **www.doverpublications.com** and see every Dover book in print. Dover publishes more than 500 books each year on science, elementary and advanced mathematics, biology, music, art, literary history, social sciences, and other areas.